ISBN 978-1-331-48161-4
PIBN 10196080

1 MONTH OF
FREE
READING

at

www.ForgottenBooks.com

By purchasing this book you are
eligible for one month membership to
ForgottenBooks.com, giving you
unlimited access to our entire
collection of over 1,000,000 titles via
our web site and mobile apps.

To claim your free month visit:

www.forgottenbooks.com/free196080

English
Français
Deutsche
Italiano
Español
Português

www.forgottenbooks.com

Mythology Photography **Fiction**
Fishing Christianity **Art** Cooking
Essays Buddhism Freemasonry
Medicine **Biology** Music **Ancient**
Egypt Evolution Carpentry Physics
Dance Geology **Mathematics** Fitness
Shakespeare **Folklore** Yoga Marketing
Confidence Immortality Biographies
Poetry **Psychology** Witchcraft
Electronics Chemistry History **Law**
Accounting **Philosophy** Anthropology
Alchemy Drama Quantum Mechanics
Atheism Sexual Health **Ancient History**
Entrepreneurship Languages Sport
Paleontology Needlework Islam
Metaphysics Investment Archaeology
Parenting Statistics Criminology
Motivational

THE
AMERICAN MUSEUM
JOURNAL

VOLUME VII, 1907

NEW YORK:

PUBLISHED BY THE

AMERICAN MUSEUM OF NATURAL HISTORY

1907

Scientific Staff

DIRECTOR
HERMON C. BUMPUS, Ph.D., Sc. D.

DEPARTMENT OF PUBLIC INSTRUCTION
Prof. ALBERT S. BICKMORE, B. S., Ph.D., LL.D., Curator Emeritus
GEORGE H. SHERWOOD, A.B., A.M., Curator

DEPARTMENT OF GEOLOGY AND INVERTEBRATE PALÆONTOLOGY
Prof. R. P. WHITFIELD, A.M., Curator
EDMUND OTIS HOVEY, A.B., Ph.D., Associate Curator

DEPARTMENT OF MAMMALOGY AND ORNITHOLOGY
Prof. J. A. ALLEN, Ph.D., Curator
FRANK M. CHAPMAN, Associate Curator

DEPARTMENT OF VERTEBRATE PALÆONTOLOGY
Prof. HENRY FAIRFIELD OSBORN, A.B., Sc.D., LL.D., D.Sc., Curator
W. D. MATTHEW, Ph.B., A.B., A.M., Ph.D., Associate Curator
O. P. HAY, A.B., A.M., Ph.D., Associate Curator of Chelonia

Prof. BASHFORD DEAN, A.B., A.M., Ph.D., Curator of Fossil Fishes
LOUIS HUSSAKOF, B. S., Ph. D., Assistant

DEPARTMENT OF ETHNOLOGY
CLARK WISSLER, A.B., A.M., Ph.D., Curator
HARLAN I. SMITH, Assistant Curator
GEORGE H. PEPPER, Assistant
CHARLES W. MEAD, Assistant

DEPARTMENT OF ARCHÆOLOGY
Prof. MARSHALL H. SAVILLE, Associate Curator

DEPARTMENT OF ENTOMOLOGY
WILLIAM BEUTENMÜLLER, Curator

DEPARTMENTS OF MINERALOGY AND CONCHOLOGY
L. P. GRATACAP, Ph.B., A.B., A.M., Curator
GEORGE F. KUNZ, A.M., Ph.D., Honorary Curator of Gems

DEPARTMENT OF BOOKS AND PUBLICATIONS
Prof. RALPH W. TOWER, A.B., A.M., Ph.D., Curator

DEPARTMENT OF INVERTEBRATE ZOÖLOGY
Prof WILLIAM MORTON WHEELER, Ph.D. Curator
ROY W. MINER, A.B., Assistant Curator
B. E. DAHLGREN, D.M.D., Assistant Curator

DEPARTMENT OF PHYSIOLOGY
Prof. RALPH W. TOWER, A.B., A.M., Ph.D., Curator

DEPARTMENT OF MAPS AND CHARTS
A. WOODWARD, Ph.D., Curator

iv

CONTENTS OF VOLUME VII.

CONTENTS.

ILLUSTRATIONS.

No. 8, DECEMBER.

LIST OF ILLUSTRATIONS.

ILLUSTRATIONS.

THE COLUMBIAN MAMMOTH

(*Elephas Columbi*)

A nearly entire skeleton, 10 feet 6 inches high at the shoulders, from near Jonesboro, Indiana. Exhibited in Hall No. 406

The American Museum Journal

Vol. VII   JANUARY, 1907   No. 1

THE present number of the AMERICAN MUSEUM JOURNAL marks a change in the policy of the magazine. In order to emphasize the news features of the periodical and make it a more satisfactory medium of communication with the Members of the Museum, the JOURNAL is to be issued monthly from October to May, inclusive, instead of quarterly, as at present. The guide leaflets will cease to form an integral part of the publication, but they will continue to be issued at intervals, and copies of them will be sent free to Members upon request made to the Director.

The guide leaflet now in press is No. 23 of the series and is entitled "Peruvian Mummies and what they Teach." The book has been prepared by Mr. C. W. Mead, assistant curator of the Department of Ethnology, and is intended as an introduction to the study of the Peruvian Hall (No. 302 of the third, or gallery, floor of the Museum). Non-members of the Museum may obtain the pamphlet for ten cents at the entrance to the building or by writing to the Librarian.

RECEPTION TO COMMANDER PEARY.

SATURDAY afternoon, December 8, by invitation of President Jesup and the members of the Peary Arctic Club Commander Peary gave in the auditorium of the Museum the first public account of his remarkable exploring expedition to the latest "farthest north." The thrilling story of the voyage of the "Roosevelt" and the dash for the Pole was told with the aid of numerous lantern slide illustrations. The personality of the speaker and the pictures added much to the vividness of the account, the bare outlines of which were already familiar to the audience through newspaper accounts. Before the lecture the Trustees of the Museum, many of the chief state and city officials, the members of the Peary Arctic Club, the officers of the American Geographical Society and the New York Historical Society, the staff of the Museum and other scientists and educators of the City and vicinity met Mr. Peary in the Board Room of the Museum by invitation of Mr. Jesup, while after the lecture Mr. Peary held an informal public reception in the auditorium.

The great desire of the people of New York to see, hear and shake hands with the intrepid explorer was manifested by the attendance at the Museum of thousands of visitors, who packed not only the auditorium but also the adjacent exhibition halls with a multitude that greeted Mr. Peary enthusiastically when he was introduced to his audience by the President and as he walked through the Museum after the lecture under the escort of the Director.

A ZOÖLOGICAL EXPEDITION TO NEW MEXICO AND ARIZONA.

URING last July and August Dr. Alexander V. Ruthven of Michigan University made a trip in the interests of the American Museum to southern New Mexico and Arizona, for the purpose of studying the reptiles and batrachians of those regions. He was accompanied by Mr. G. von Krochow of the Museum staff. The territory about Alamogordo, New Mexico, and Tuscon, Arizona, was explored in detail. In each of these localities there is a variety of habitats, ranging from the lower, arid deserts to the more humid, forest-crowned summits of the adjacent mountain ranges. Field stations under different environmental conditions were established, where, after noting the physical factors and vegetation of each habitat, the reptile and amphibian life was determined by careful collecting.

As a result of this work it was established that the reptile and amphibian life of each habitat was characteristic, as has been shown for some other animals and for plants. Notes were made on the habits of the different forms to determine their relationship to the environment with which they have been found associated, observations on food habits being supplemented by examinations of the contents of many stomachs, which have been carefully labeled and preserved.

About 1,000 specimens of cold-blooded vertebrates were secured for the Museum, besides a small collection of invertebrates, consisting principally of ants, molluscs and parasitic worms. Several valuable specimens of lizards are represented in the collection by extensive series. The results of this expedition are now being worked up in several papers, the molluscs by Mr. Bryant Walker, the vertebrates by Dr. Ruthven.

THE SKELETON OF THE COLUMBIAN MAMMOTH.

NE of the most recent additions to the collections in the Hall of Vertebrate Palæontology is a skeleton of the Mammoth, *Elephas columbi*. This skeleton shows that the Mammoth had a very short back and long legs. Its body was not as massive as that of the Mastodon, and the pelvis is proportionately narrower. The head was carried more erect than that of the Mastodon, and the tusks, which point down at first, then curve forward and upward and completely cross at their points. The tusks of the Mastodon, on the other hand, continue farther downward before bending upward, and their distal portions turn outward.

During the middle Pleistocene or Glacial Period three well-known species of Mammoth inhabited North America. The Siberian mammoth, *Elephas primigenius,* was abundant in the northern part of America, and bodies of animals of this species have been found in the ice and frozen ground of Alaska, the flesh and hair still preserved intact. The Imperial Mammoth, *Elephas imperator,* inhabited the southern United States and its remains are found in Texas. The third species, the Columbian Mammoth, *Elephas columbi,* inhabited the greater part of the United States, and its range extended as far southwest as Mexico.

Teeth and bones of the last-named species have been identified from many localities, but the specimen now under consideration is the most nearly complete skeleton yet mounted in this country. This skeleton was found on the farm of D. C. Gift, four miles east of Jonesboro, Indiana. That part of Grant County is level and was originally swampy and had to be drained. While enlarging a drainage canal across a part of the farm, a tenant encountered the skeleton only eight feet below the surface, where it lay articulated, with its bones in position just as the animal had fallen after becoming mired in the old swamp. The feet were not found, having been perhaps scraped out and lost during the first opening of the drain, hence the lower parts of the limbs and the feet have been restored from casts of a skeleton of *Elephas primigenius* which is in the Paris Museum of Natural History.

The following measurements show the size of this unique specimen:

Length, tips of tusks to vertical line of tail, 17 feet 9½ inches.

"   base "   "   "   "   " " " 13 " 3½ "

Height at shoulders,   10 " 6   "

Length of right tusk (outside curve),   11 " 4½ "

DEPARTMENT OF VERTEBRATE PALÆONTOLOGY; FIELD EXPEDITIONS OF 1906.

HE Department of Vertebrate Palæontology had three expeditions in the field last summer. The first, under Mr. Barnum Brown, continued the hunt for dinosaurs in the Upper Cretaceous deposits of Montana. A fine skeleton of the Duck-billed Dinosaur *Claosaurus* and several less complete specimens of these strange looking animals were secured. The Museum now possesses two mountable skeletons and an excellent skull, representing three distinct kinds of Duck-billed Dinosaurs, besides many less complete specimens. Mr. Brown also discovered remains of a large dinosaur which appears to be entirely new to science and of a small species related to the Bird-catching Dinosaur of the Jurassic period. The remarkable variety of dinosaurs from the Upper Cretaceous, most of them of huge size and bizarre appearance, is a great incentive towards continuing the search for them with especial vigor. It is hoped that the final results will enable the Museum to fill an entire exhibition hall with the dinosaurs of this period. In addition to the dinosaur remains a few specimens of the minute and very rare Upper Cretaceous mammals were secured.

The second expedition, under Mr. Walter Granger, continued the exploration of the Eocene formations of Wyoming, working this year principally in the Washakie Basin. A number of fine specimens were obtained of the characteristic large mammals of this formation, the *Eobasileus*, largest and last of the uintathere race, *Amynodon*, the first of the rhinoceroses, *Acharnodon*, a gigantic pig-like animal, *Patriofelis*, a large and powerful carnivore, and of the Eocene ancestors of the Titanotheres, besides skulls and skeletons of several of the smaller carnivora, rodents and other forms. Many of these will be new to science and all of them will aid greatly in filling out our collections from the later Eocene. Mr. Granger paid much attention to the stratigraphy of the formation, and the origin of the materials composing it, obtaining data which, with the collections of fossils secured at different levels, will enable us to fix accurately the time relations of the Washakie to the Bridger and Uinta formations (Middle and Upper Eocene).

SCENE IN WASHAKIE BASIN

The third expedition was conducted by Mr. Albert Thomson in the Lower Miocene of South Dakota. The formations of this age in the Western States are in general very barren of fossil mammals and have been but little explored, although the formations above and beneath them have yielded large collections to the explorations of the past half-century. By dint of diligent and thorough search Mr. Thomson succeeded in getting from these unpromising beds a considerable collection of skulls and skeletons, nearly all of them new to science and representing an intermediate stage between the Oligocene and later Miocene.

Among the interesting novelties of this collection is a fragmentary skeleton of a gigantic Wolverene as large as a jaguar or a black bear.

It was much the largest of the weasel family and was a truly formidable beast of prey, if it possessed the savage and bloodthirsty disposition of its modern relatives.

Professor Osborn, under whose directions the expeditions were sent out, visited all the parties during the summer and gathered valuable data for his studies upon the stratigraphy and relative age of the Tertiary formations. Dr. Matthew was with Mr. Thomson's expedition during the first half of the summer, and Mr. Gregory was of the party in the latter part of the season.

THE SELMA METEORITE.

HE collection of meteorites in the foyer of the Museum has recently been enriched by the addition of an aërolite, or stone meteorite, which was found in March, 1906, about two miles north-northwest of Selma, Alabama, near the road to Summerfield. The fortunate finder was Mr. J. W. Coleman of that city.

Mr. Coleman states his belief that the meteorite fell on July 20, 1898. At about 9 o'clock of the evening of that day at least five observers at different stations from half a mile to two and one half miles from where the stone was found saw a brilliant meteor pass through the air leaving a "trail of fire ten or twelve feet long." The meteor seems to have traveled in a direction somewhat west of north, and its flight is said to have been accompanied by a heavy, rumbling noise. No other similar meteorite has been noted in the immediate region, and this meteorite does not seem to show any more decomposition of surface than might have taken place in the eight years that have elapsed since the date of its assumed fall. The identity of this find with the shooting star of July 20, 1898, cannot of course be established with certainty, but it seems probable.

This meteorite, photographs of which are reproduced in this number of the JOURNAL, is one of the ten largest aërolites ever found. Most such bodies break to pieces in the earth's atmosphere, probably on account of unequal heating due to friction against the air, and shatter into scores and even hundreds or thousands of fragments before they reach the ground, and this is the largest entire aërolite now in the United States.

THE SELMA METEORITE

Weight, 306 lbs.   Front or " Brustseite."

The Selma, as this meteorite will be called, is 20½ inches high, 20 inches wide and 14 inches thick, and it weighs 306 pounds (138.6 kilos). A piece of perhaps four pounds weight has been lost from the mass,

THE SELMA METEORITE
Rear View.

hence it is probable that the original weight was 310 pounds. It has lain buried in the ground where it fell for several years, so that the original glassy crust has been largely decomposed and washed away, and the characteristic "thumb-mark" pittings have been partly obscured. Some portions, however, remain as an indication of its original condi-

THE SELMA METEORITE
Views of opposite edges.

tion. In shape the meteorite is roughly polygonal, without pronounced orientation features, though it seems probable that the side shown on page 9 was the "brustseite," or front, during flight through the atmosphere. The mass is deeply penetrated by cracks on both sides, but principally on the rear. The cracks do not radiate from one or more centers, nor is the apparent rigidity of the mass affected by them; hence they do not seem to have been caused by shattering due to impact with the earth. The position and character of the fissures indicate that they were due to unequal heating through friction with the air during flight through the atmosphere, the tension produced being insufficient to cause complete fracture.

Macroscopic examination of a cut and polished fragment shows the stone to have a dark brownish-gray color, and to be made up of spherical or nearly spherical "chondrules," or particles, firmly imbedded in a similar matrix. The largest chondrules are $\frac{1}{8}$ inch (3 mm) in diameter, though those more than one half as large are rare. A strong magnifying glass is needed to show one the minute grains of iron scattered through the mass.

The specific gravity of the stone is 3.42 as determined upon a fragment weighing 4.56 ounces (129.4 g) and showing some effects of decomposition. A chemical analysis of the material has not yet been made, but Dr. G. P. Merrill of the National Museum has had sections cut and polished and has published a brief scientific description of the meteorite in the Proceedings of the U. S. National Museum for 1906, where he gives the find the name which we have adopted.

E. O. HOVEY.

MUSEUM NEWS NOTES.

At the autumn meeting of the Board of Trustees President Jesup reported that since the last meeting of the Board important gifts had been received, as follows :

From Mr. ARTHUR CURTISS JAMES, for mural decorations in the Eskimo Hall, and for field explorations for fossil horses;

From Mr. GEORGE S. BOWDOIN, for the development of the Cetacea Collection;

From Mr. SAMUEL V. HOFFMAN, for an entomological expedition to the Black Mountains of North Carolina;

From Mr. JAMES R. KEENE, the skeleton of his famous race horse
"Sysonby;"

From Mrs. F. K. STURGIS an additional contribution for the purchase of
Birds of Paradise;

From Mr. FRANCIS R. APPLETON for exploration for fossil horses;

From Mr. A. C. CLARKSON, the library and entomological collection of
the late Mr. Frederick Clarkson.

PROFESSORS J. A. ALLEN AND HENRY F. OSBORN represented the
Museum at the autumn meeting of the National Academy of Sciences
which was held in Boston on November 20-22 in the new buildings of
the Harvard Medical School. At this meeting Professor Osborn presented two papers, one on "Section of American Tertiaries," and the
other on "A Complete Mounted Skeleton of the Finback Lizard *Naosaurus* of the Permian." Among the extensive series of exhibits displayed at the conversazione held in connection with the meeting there
were the following by Professor Osborn: (1) Recent restorations of
extinct horses of North America, executed by Charles R. Knight, under
direction of Henry F. Osborn, water-colors, photographs; (2) first complete section of the American Tertiaries,— a preliminary study.

UPON receipt of the telegrams announcing the "farthest north"
attained by Commander Robert E. Peary, U. S. N., the Arctic explorer, a temporary exhibit was installed pertaining to Mr. Peary's
Arctic work. One of his sledges harnessed to four mounted specimens
of the dogs which he used on one of his earlier expeditions together with
a figure dressed in Arctic clothing showed the means of transportation
employed by the explorer. On the sledge was one of the sleeping-bags
so essential to existence in the Arctic regions. On a neighboring table
were shown photographs of Peary's steamer, the "Roosevelt," together
with photographs of the explorer and his party. The question as to
how Peary determined his position was answered by the display of a
sextant for determining latitude and a chronometer for determining
longitude used by General Greely in his famous polar work and abandoned by him at Fort Conger. On a section of a globe nearby illustrating the north polar regions, Peary's route to the farthest north position
was shown by means of colored cord.

AMONG recent acquisitions to the Department of Mineralogy special
mention should be made of a large series of nuggets and coarse gravel

Gold from placer diggings in Alaska aggregating about 4 pounds troy in weight; a wonderful crystal and a magnificent group of ruby-red Tourmaline (Rubellite) of gem quality and deep color from the noted locality at Pala near San Diego, California; and a block of quartzite from New South Wales bearing a surface of more than 90 square inches of Precious Opal.

TUESDAY, October 30, the National Association of Audubon Societies for the Protection of Wild Birds and Animals held its annual meeting in the West Assembly Hall.

THE Museum recently acquired by purchase a fine collection from the Samoan Islands. Among other things this contains a complete outfit for the manufacture of bark or "tapa" cloth. In the manufacture of this cloth, single strips of bark from a species of mulberry are prepared by scraping and soaking in water, after which they are beaten out very thin by means of small wooden clubs. These thin sheets, while still wet, are laid one over another and the whole beaten together to form a large sheet of uniform thickness. Such bark cloth is in some respects a kind of paper, but it is serviceable as cloth since it is not easily damaged by water. The finished cloth is often ornamented by printing, or rather rubbing. For this purpose designs in relief are carved on wood or built up of palm-leaf cuttings, upon which the cloth is laid and rubbed with sticks of coloring matter, like crayon. This leaves an impression of the raised portion of the carving similar to that produced when a school-boy rubs the impression of a coin into the fly-leaf of his book. Aside from the tapa outfit, the collection contains several handsome pieces of finished cloth and a number of costumes, household utensils and other implements. It is proposed to install this collection together with other material in the Museum as a special exhibit from one of the colonial possessions of the United States.

DR. E. O. HOVEY of the Department of Geology returned in the latter part of October from the convention in Mexico of the Tenth International Geological Congress to which he was sent as the delegate of the Museum. The convention met in Mexico City September 6 to 14, inclusive, and was attended by many of the prominent geologists of Europe as well as of the United States. Dr. Hovey brought back with him a large series of specimens of ores, rocks and fossils and many photographs (negatives) for the Museum collections.

A valuable collection of the co-types of the ants obtained by Professor Filippi Sylvestri of Naples in a trip across South America some years ago has been presented to the Museum by the collector. The corresponding type collection was described by Professor Carlo Emery of Bologna.

The great Scudder collection of fossil ants from Florissant, Colorado, has been loaned to the Museum for study and description. The collection consists of more than 4,000 specimens representing about 40 species. None of these species has been described, and there is ample material for the work. The specimens are almost entirely males and females which dropped into the Tertiary lake of Florissant while on their nuptial flight. Practically no workers are represented in the series.

LECTURES.

Legal Holiday Course.

Upon the four principal legal holidays occurring during the winter season the Museum has for many years given lectures free to the public, no tickets being required for admission. The programme for the current season follows. The doors open at 2:45 and the lectures begin at 3:15.

Thanksgiving Day, November 29, 1906.— E. O. Hovey, "Volcanoes,"

Christmas Day, December 25, 1906.—Louis P. Gratacap, "Iceland: Its Scenery and Inhabitants,"

New Year's Day, January 1, 1907.— Frank M. Chapman, "The Home-Life of Flamingoes,"

Washington's Birthday, February 22, 1907.— Harlan I. Smith, "The Five American Nations: Conquerors of the Snow, Forest, Mist, Desert and Plain."

People's Course.

The programme of illustrated Free Lectures to the People for January is as follows:

Tuesday evenings at 8 o'clock.

January 8.— E. C. Culver. "The Yellowstone National Park."

January 15.— Colvin B. Brown. "The Sierra Nevada Mountains and the Yosemite valley."

January 22.— J. W. Fairbank. "Ramona and the Land of Sunshine."

January 29.— George Wharton James. "Primitive Inventions. What We Owe to the Indian Inventor."

February 5.— "The Religion of the Southwest Indians."

Saturday evenings at 8 o'clock.

A course of lectures by Professor H. E. Crampton, Columbia University.

January 5.— "Principles of Organic Evolution."

January 12.— "The Anatomical Evidence of Evolution."

January 19.— "Development as Evidence of Evolution."

January 26.— "The Evidence of Fossils."

These lectures are given in coöperation with the Department of Education of the City of New York. They are open free to the public and no tickets are required for admittance, except in the case of children, who, on account of the regulations of the Department of Education, will be admitted only on presentation of the ticket of a Member of the Museum.

The doors open at 7:30 o'clock and close when the lectures begin.

MEETINGS OF SOCIETIES.

Meetings of the New York Academy of Sciences and Affiliated Societies are held at the Museum according to the following schedule:

On Monday evenings, The New York Academy of Sciences:

First Mondays, Section of Geology and Mineralogy.

Second Mondays, Section of Biology.

Third Mondays, Section of Astronomy, Physics and Chemistry.

Fourth Mondays, Section of Anthropology and Psychology.

On Tuesday evenings, as announced:

The Linnæan Society, The New York Entomological Society and the Torrey Botanical Club.

On Wednesday evenings, as announced:

The New York Mineralogical Club.

The programme of meetings of the respective organizations is issued in the weekly "Bulletin" of the New York Academy of Sciences and sent to the members of the several societies. Members of the Museum on making request of the Director will be provided with these circulars as they are published.

The following lectures have been given at the Museum recently through coöperation with the Academy:

October 29.— Dr. F. A. Lucas, " Elephants, Past and Present."

November 19.— Professor C. L. Poor, "The Proposed Astronomical and Nautical Museum for New York City."

December 5.— Charles Truax, "The Yellowstone National Park."

NORTHEAST QUARTER OF FOYER

Showing five of the busts of American Men of Science

18

The American Museum Journal

| Vol. VII | FEBRUARY, 1907 | No. 2 |

MEMORIALS OF MEN OF SCIENCE.

ATURDAY afternoon, December 29, the impressive unveiling ceremonies of the busts which have been installed in the foyer representing ten of the men who have been foremost in the advancement of science in America were held in the large auditorium. After singing the national anthem "America," Doctor Hermon C. Bumpus, Director of the Museum, acting for Mr. Morris K. Jesup, President, presented the busts to the Trustees. After the gift had been accepted by the Hon. Joseph H. Choate on behalf of the board, memorial addresses were delivered in accordance with the following programme:

Benjamin Franklin by Dr. S. Weir Mitchell of Philadelphia; Alexander von Humboldt, by His Excellency Baron Speck von Sternburg, German Ambassador, (Read by Major T. von Körner, Military Attaché of the Embassy); John James Audubon, by Dr. C. Hart Merriam, Chief, U. S. Biological Survey, Washington, D. C.; John Torrey, by Dr. Nathaniel L. Britton, Director, New York Botanical Garden, New York City; Joseph Henry, by Dr. Robert S. Woodward, President, Carnegie Institution, Washington, D. C.; Louis Agassiz, a letter was read from the Rev. Edward Everett Hale, an intimate personal friend of Professor Agassiz; James Dwight Dana, by Dr. Arthur Twining Hadley, President, Yale University, New Haven, Conn.; Spencer Fullerton Baird, by Dr. Hugh M. Smith, Deputy Commissioner, Bureau of Fisheries, Washington, D. C.; Joseph Leidy, by Dr. William Keith Brooks, Johns Hopkins University, Baltimore, Md.; Edward Drinker Cope, by Dr. Henry Fairfield Osborn, Curator, Department of Vertebrate Palæontology, American Museum of Natural History.

The occasion was a notable event in the annals of the Museum and of American science in general. The auditorium was crowded to its full capacity with members of the Museum and visiting scientists, and hundreds were turned away from the hall for lack of space for their accommodation. The addresses were alive with interest and some of them were historic in value for their appreciation of great men of the

WILLIAM COUPER

The sculptor of the busts in the Foyer

past by great men of the present who were their pupils or their friends. A pamphlet containing the addresses in full and illustrated with photographs of the busts is now being prepared to serve as a worthy memento of the occasion and a guide leaflet to the foyer.

The artist who prepared the busts is Mr. William Couper of this city, a sculptor of international reputation. The busts are intended to represent the subjects in the prime of life and at the zenith of their powers. The data used were photographs, painted and other portraits, descriptions by contemporaries and the criticisms and suggestions of friends and relatives, as far as obtainable.

EXHIBITION OF THE PROGRESS OF SCIENCE.

S a feature of the great convention of American scientists which was held from December 27 to January 2 at the Museum, Columbia University and other institutions in the city, the New York Academy of Sciences in coöperation with the Museum assembled material from all over the country to represent the progress in science, both pure and applied, that has been made during the past few years. It would be impracticable to give, in the space available in the AMERICAN MUSEUM JOURNAL, an exhaustive account of the noteworthy features of the exhibition and we shall attempt only to touch upon a few of the most salient points. The exhibition was given Friday and Saturday, December 28 and 29, under the direct auspices of the Academy and was continued under the auspices of the Museum for two weeks longer.

In the department of Anatomy a series of excellent preparations of electric fishes and dissections and diagrams of their electric organs was surprising as showing the number of such species and the strength of current induced by their peculiar organs. The exhibit of Astronomy was declared by an eminent authority to be the most remarkable and important assemblage of its kind thus far gotten together in this country, since it showed all the most important recent work of all the great observatories in the United States. In the department of Bacteriology there were extensive exhibits of cultures and methods from all the great institutions in this city which are carrying on researches in this science, including the American Museum. From the point of view of the specialist, the exhibition was most interesting, but it required too much explanation for the average visitor to spend the time over it that it deserved.

The department of Botany contained many features of general interest, among which may be mentioned the natural Spineless Cactus from the Island of Culebra, Porto Rico; a set of more than 100 dairy fungi and bacteria, including those that give the flavor to Camembert cheese, produce the sweet and sour curdling of milk and turn milk different colors; and a series of remarkable photomicrographs of thin sections of American woods.

The Chemistry exhibit was large. Its most popular features, perhaps, were those pertaining to radium and the effects of its use in surgery. The department of Electricity was much in evidence at the exhibition, on account, particularly, of the Telharmonium and the "Helion Filament" incandescent lamps. The former is a newly devised musical instrument, while the latter is a new lamp that bids fair to introduce a revolution in methods of illumination. The production of the helion lamp follows closely upon the successful manufacture of the tantalum and the tungsten lamps and marks the third wonderful discovery in incandescent lighting within two years. The new filament consists of silicon deposited upon carbon, and, although it costs more than the ordinary carbon filament, it lasts much longer and gives more than three times as much light for the same consumption of power.

In the department of Ethnology and Archæology local interest centered around a large earthenware jar of Iroquoian Indian manufacture which was exhumed recently at 214th Street and 10th Avenue, Manhattan Island. The specimen is not unique, but it is the most perfect that has been found in this vicinity. An exhibit having important bearing on the antiquity of man in America was that of some worked bone objects from caves in California. The objects are apparently of Glacial or interglacial age.

The department of Experimental Evolution attracted the attention of the public through the crowing of some live roosters introduced for the purpose of showing the effects of breeding along certain lines. Other interesting exhibits were specimens and diagrams showing the manifestations of the Mendelian law of inherited characteristics under cross-breeding. In the department of Experimental Psychology mention may be made of a series of photographs of the movements of the eye in examining sundry objects and particularly in reading, together with the apparatus used in making the negatives.

Geology and Geography were well represented, particularly through the maps and other publications showing the recent work of the United States and several State geological surveys. Special mention should be made of the new unpublished maps of Connecticut and Massachusetts. The American Museum contributed to this department models of Martinique and Mt. Pelé and a series of transparencies of the West Indies and Mexico.

In the section of Mineralogy the most important displays of popular interest were the series of specimens of unusually large and perfect crystals of Calcite collected last summer in the northern part of New York State by an expedition from the State Museum and the set of great ruby red Tourmalines and Beryls from near San Diego, California. The American Museum showed here some of the most striking recent accessions to its cabinet including a particularly handsome group of large crystals of Ruby Tourmaline (Rubellite) from California.

Invertebrate Palæontology does not ordinarily contribute showy material to an exhibition, but attention was attracted at the Academy "conversazione" by the display of Trilobites, Eurypterids (crustaceans) Hexactinellids (glass sponges) and other fossils sent down by the New York State Museum from recent collecting in the central and western part of the state. The American Museum contributed to this section some remarkable Cretaceous cephalopods and a series illustrating the Jurassic fauna of the Black Hills of South Dakota and Wyoming.

The major portion of the display in the section of Vertebrate Palæontology was contributed by the American Museum and consisted of many remarkable specimens most of which have been noticed in previous issues of the JOURNAL. Mention should be made, however, of the skeleton of the strange-looking Fin-backed Lizard, *Naosaurus*, the restorations of several fossil fishes and the practically complete specimen of a giant Tortoise from the Badlands of Wyoming. This Tortoise has its nearest living relatives among the big turtles of the Galapagos Islands.

To the sections of Pathology, Pharmacology, Physics and Physiology the American Museum made no contributions of exhibits. The section of Physics, as usual, attracted much attention from visitors.

Recent earthquakes aroused much interest in a new form of seismograph and its records for New York City. A series of photographs taken with the lens from a fish's eye was worth more than a passing glance. The Zeiss epidiascope is a new and successful electric lantern device for throwing on a screen pictures from ordinary photographic prints and other opaque objects. A prominent feature of the Physics exhibit was a lecture Saturday evening, December 29, by Professor E. F. Nichols on "The Pressure of Light," illustrated with experiments and lantern slides. The Physiology exhibit was extensive and contained much of technical as well as popular interest. Something in motion always attracts a crowd, hence the popularity of the zoëtrope showing a series of kinetoscope pictures of the movements of the alimentary canal taken with X-ray apparatus, and the working model of the heart showing perfectly the action of valves.

By far the most attractive exhibit of all, and the one showing most clearly the advances made during the past five years, at least along certain lines, was that of the department of Zoölogy. The exhibit consisted principally of contributions from the American Museum of Natural History, the New York Zoölogical Park, the Aquarium and the Brooklyn Museum. Among the American Museum exhibits, most attention was attracted perhaps by that of two mounted lions, showing in most striking manner the advance that has been made in a comparatively few years over the old methods of taxidermy. The new ideas were exemplified in the mount of " Hannibal," the lion, so well known to many visitors at the Zoölogical Park in Bronx Park. Other exhibits were entered by the Museum, but not removed from the exhibition halls. These were the group of Wapiti, or American Elk, in the East Mammal Hall, second floor of the building, and ten groups of North American birds with panoramic backgrounds, illustrating a novel method of lighting. In seven of these the background is curved, so as to represent the horizon in its natural position. Mention should be made of the wax-and-glass models of various invertebrates, particularly some insects, the familiar Squid of the New England coast and an Actinian (sea anemone). The exhibit from the Zoölogical Park consisted, for the most part, of maps, photographs, transparencies and samples of labels. The New York Aquarium, in addition to similar exhibits, showed some of the marine aquaria which are supplied to the public schools of the city. The Brooklyn Institute Museum exhibited typical collections showing its efforts to disseminate knowledge outside the Museum. Japanese Sharks, Hag Fishes and other curious marine forms were very interesting to the specialist, and Economic Entomology was illustrated, in part, by the photographs, transparencies and specimens illustrating recent progress in New Jersey in the extermination of the mosquito and in methods of educating the public on the subject.

EXPEDITION TO THE DESERT OF FAYOUM, EGYPT.

N January 5 Professor Henry F. Osborn sailed for Egypt accompanied by Messrs. Walter Granger and George Olsen of the Department of Vertebrate Palæontology on an exploring expedition of three months into the Fayoum desert. In 1900 Professor Osborn[1] predicted that the remote ancestors of the Proboscidea, Sirenia and Hyracoidea would prove to be of African origin, and soon afterward, through the extensive exploration and study of this region by the Egyptian Survey, this prophecy came true. This desert has yielded some of the most remarkable recent discoveries in palæontology, among which may be cited, besides those in the three orders above mentioned, many entirely new and unique forms, one of which is *Arsinoitherium*. Dr. C. W. Andrews of the British Museum and Mr. Hugh J. N. Beadnell of the Egyptian Survey have been the principal students of this fauna and have described their discoveries in a series of papers published during the last five years, culminating in a large quarto memoir published last year by Dr. Andrews.

Ever since the fulfillment of his prophecy and the discovery of this fauna, new to science, Professor Osborn has been anxious to visit and explore the Fayoum, but he felt that he could not go before the publication of Dr. Andrews's report freed the field to all scientific workers. Palæontologists at present regard Africa as the storm center of their work and look to the revelation of its secrets for the solution of many of the problems which confront them in the unraveling of the past. If the expedition is successful, the addition of this fauna to the collection of fossil vertebrates in our Museum will greatly enhance its interest to the public and its value to the student.

[1] Faunal Relations of Europe and America during the Tertiary Period and Theory of the Successive Invasion of an African Fauna into Europe. Ann. N. Y. Acad. Sci., Vol. XIII, No. 1, July 21, 1900, pp. 1–72.

President Jesup provided the necessary pecuniary support for the ·expedition, and Professor Osborn embarked on his undertaking with letters from President Roosevelt to Lord Cromer, the head of the Egyptian government, introducing him as Geologist and Palæontologist of the U. S. Geological Survey, as well as Vice-president and Curator of this Museum, and from eminent English and German scientists who have preceded him in the work of exploring this desert. Professor Osborn will return about April 1 and will doubtless have a story of experiences very different from those incident to the exploration of our own western country, to which he has devoted so much time in the last fifteen years, interesting as the latter have been. The party goes by camel train southwestward from Cairo three days' journey into the desert, and the region in which work is to be carried on is at least fifty miles from the nearest source of water.

THE MUSEUM "BULLETIN" FOR 1906.

HE twenty-second volume of the BULLETIN of the American Museum was issued during the year 1906. The articles, which pertain to the scientific work of the museum and are technical in character, are also published separately and, like the volume itself, may be obtained from the Librarian. The table of contents of the volume is as follows:

MUSEUM NEWS NOTES.

PROFESSOR William M. Wheeler, Curator of Invertebrate Zoölogy, lectured in the auditorium on "The Polymorphism of Social Insects" before a large audience on Friday, December 28. The lecture was followed by a meeting at which was organized the Society of American Entomologists, with Professor J. H. Comstock of Cornell University as president and Doctor E. S. G. Titus of Washington, D. C., as secretary. The society includes economic as well as other working entomologists and begins life with nearly three hundred members. In connection with the meeting an interesting exhibition was made, the principal features of which were some air-brush enlargements of sundry insects for lecture room purposes, a berlese apparatus for catching insects, a wonderful series of photographs of spiders' webs, an artificial ant nest and a series of excellent specimens of fossil insects from Florissant, Colorado.

ON December 28 and 29 the Geological Society of America held largely attended sessions of its annual meeting at the Museum. This meeting formed a portion of the great convention of scientific societies which was held in New York City from December 27 to January 3, inclusive.

On Saturday evening, December 29, a reception to the visiting delegates of the American Association for the Advancement of Sciences and affiliated societies was given at the Museum by the trustees in co-operation with the New York Academy of Sciences. One of the features of this reception was a series of brief lectures with lantern slide illustrations which were given in the large lecture hall according to the following programme:

"Mt. Pelé and St. Pierre," by E. O. Hovey;

"The Fur-Seal Islands of Alaska," by Charles H. Townsend;

"The Home-Life of the Brown Pelican," by F. M. Chapman;

"The Fire-Walking Ceremony of Tahiti," by H. E. Crampton;

"Illustrations of Wild Flowers used in Lectures by the Society for the Preservation of Native Plants," by Charles L. Pollard;

"Recent Explorations and Results of the Department of Vertebrate Palæontology," by Henry F. Osborn.

THE American Bison Society held its annual meeting at the Museum on Thursday, January 10. This society has for its object not only the prevention of the extermination of the Bison, but also the encouragement of the raising of the animal as a commercial proposition. A generation ago the Bison, or American Buffalo, roamed over the western plains in vast herds, estimated to contain more than ten million individuals, while to-day, on account of the merciless and wanton slaughter practised in the early eighties, scarcely two thousand are known to be in existence. The society proposes to encourage the establishment of Bison reservations in each state where climate and other conditions are favorable for the maintenance and increase of herds. For New York the proposition is that, as a beginning, the State set aside nine square miles in one of the reserved areas of the Adirondack region and appropriate $15,000 for the purchase and maintenance of a herd of fifteen Bison. Dr. William T. Hornaday, director of the New York Zoölogical Park, is the president of the society.

THE Department of Vertebrate Palæontology has added to its Horse Alcove, through the generosity of Mr. Randolph Huntington, the skeleton of the Arabian stallion "Nimr." The skeleton has been mounted by Mr. S. H. Chubb under the direction of Professor Osborn in such a manner as to show the characteristics of the Arabian Horse, particularly the high elevation of head and tail when the animal is excited. This race of Horse is characterized by a small skull with high, prominent, broadly separated orbits, slender nose and concave profile, horizontal position of pelvis, round thorax, long and slender cannon bones and pasterns. The mounting of the skeleton was undertaken only after long study of the Arabian horses in the Huntington stud and many photographs taken from life.

MISS MARY LOIS KISSELL has been engaged by the Department of Ethnology to arrange the exhibits of basketry for the various North American tribes. The Museum possesses large collections of some very rare Californian baskets, particularly the so-called "ti-stitch" of the Pomo. The entire Pomo collection has been re-arranged according to the weave, and labels for the specimens are being prepared. The plan of the new exhibit includes a general synoptic series for the chief weaves employed in the different parts of the world. Following this it is proposed to arrange according to locality the Museum's large collection of baskets from California and the adjacent parts of the Pacific Coast.

THE American Institute of Social Service opened an Exposition of Safety Devices and Industrial Hygiene at the Museum Tuesday, January 29, to continue two weeks. The exhibition, which is free to the public, comprises live machinery, working models and photographs from various European countries, as well as from the United States, and shows in striking manner the fact that we are nearly a generation behind Europe in any organized effort to protect workmen from injury while they are at work. The exhibition has aroused much public interest on account of the great recent increase in accidents affecting the life, limbs and health of American workmen.

THE record of attendance at the Museum during the year 1906 was 476,133 visitors.

LECTURE ANNOUNCEMENTS.

MEMBERS' COURSE.

THE second course of lectures for the season 1906–1907 to Members. of the American Museum of Natural History and persons holding complimentary tickets given them by Members will be given during February and March. The lectures will be delivered on Thursday evenings at 8:15· and will be fully illustrated by stereopticon views. The programme for the course is as follows:

February 21.— FRANK M. CHAPMAN, "The Birds of Spring."
February 28.— RICHARD TJÄDER, "Hunting Big Game in British East Africa."
March   7·— FREDERIC A. LUCAS, "Whales and Whaling."
March   14.— EDMUND OTIS HOVEY, "Earthquakes; Their Causes and Effects."
March   21.— CLARK WISSLER, "Living with the Indians of the Plains."

PUPILS' COURSE.

THE Lectures to Public School Children will be resumed in March and will be given in accordance with the following programme.

	Mar.	Apr.	
Monday,	4	8.—"Along the Historic Hudson." By G. H. Sher-- wood.
Wednesday,	6	10.—"Life in the Far North." By H. I. Smith.
Friday,	8	12.—"New York City in Colonial Days." By R. W.. Miner.
Monday,	11	15.—"The American Indians of today." By G. H.. Pepper.
Wednesday,	13	17.—"Commercial Centers of Europe." By E. O.. Hovey.
Friday,	15	19.—"Farming and Ranching in the United States."· By G. H. Sherwood.
Monday,	18	22.—"Travels in South America." By Barnum Brown.
Wednesday,	20	24.—"Natural Wonders of our Country." By R. W. Miner.
	Apr.	
Friday,	5	26.—"The Products of Our Mines." By E. O. Hovey.

These lectures are open to public school children accompanied by their· teachers and to the children of Members of the Museum on the presentation of membership tickets. Particulars of this course may be learned by· addressing the Director of the Museum.

PEOPLE'S COURSE.

Tuesdays at 8 p. m.   The continuation of a course of illustrated lectures by George Wharton James.

February 5.— "The Religion of the Southwest Indians."

February 12.— "The Prehistoric and Aboriginal Dwellers of the Southwest."

February 19.— "The Colorado Desert: Its Horrors, Mystery and Reclamation."

February 26.— "Things We May Learn from the Indians."

Saturdays at 8 p. m.   The continuation of a course on Evolution by Professor H. E. Crampton of Columbia University.

February 2.— "The Method of Evolution."

February 9.— "The Evolution of the Human Species."

February 16.— "The Evolution of Human Races."

February 23.— "Evolution of Mind, of Society and of Ethics."

These lectures are given in coöperation with the Department of Education of the City of New York.   They are open free to the public and no tickets are required for admittance, except in the case of children, who, on account of the regulations of the Department of Education, will be admitted only on presentation of the ticket of a Member of the Museum.

The doors open at 7:30 o'clock and close when the lectures begin.

MEETINGS OF SOCIETIES.

Meetings of the New York Academy of Sciences and Affiliated Societies are held at the Museum according to the following schedule:

On Monday evenings, The New York Academy of Sciences:

First Mondays, Section of Geology and Mineralogy.

Second Mondays, Section of Biology.

Third Mondays, Section of Astronomy, Physics and Chemistry.

Fourth Mondays, Section of Anthropology and Psychology.

On Tuesday evenings, as announced:

The Linnæan Society, The New York Entomological Society and the Torrey Botanical Club.

On Wednesday evenings, as announced:

The New York Mineralogical Club.

The programme of meetings of the respective organizations is issued in the weekly "Bulletin" of the New York Academy of Sciences and sent to the members of the several societies.   Members of the Museum on making request of the Director will be provided with these circulars as they are published.

THE BARBARY LION "HANNIBAL"

The American Museum Journal

| Vol. VII | MARCH, 1907 | No. 3 |

THE AFRICAN LION "HANNIBAL."

O much favorable comment has come to the Museum regarding the illustration of the mounted lion that was issued with the announcement of the spring courses of lectures that we think that our Members will be interested in knowing how the specimen was acquired and how it was prepared and mounted.

On October 17, 1902, Miss Carnegie, daughter of Andrew Carnegie, presented to the New York Zoölogical Park an excellent example of the Barbary lion. On February 21, 1905, "Hannibal," as the lion was called, died, and the body was presented to the American Museum, through the courtesy of the Zoölogical Society. About a year later, Mr. James L. Clark, the Museum's animal sculptor, began preparations for modeling the animal. The work was completed a few weeks ago, and the lion was seen by the public for the first time at the Exhibition of the New York Academy of Sciences which was held at the Museum from December 27, 1906, to January 14, 1907. It will soon be placed on permanent exhibition.

When Mr. Clark began preparations for mounting the lion he visited the Zoölogical Park and made a study in clay from living specimens. This was prepared with great care, attention being paid to every detail of structure. After the small model had been completed, the real work of mounting began. The general outline of the animal was obtained, and the basis of the life-sized model formed, exactly as a sculptor makes an armature for a large figure. On this foundation wet clay was piled until the mass corresponded in some degree to the measurements which had been made from Hannibal in the flesh. Modeling tools in trained hands then developed the surface and reproduced with precision the contours of muscle, cord and tendon.

From time to time the skin was placed over the clay to insure an exact fit, and any imperfections in the model were corrected. When at last the desired form had been attained, a plaster mold was taken, from which a cast was made. This cast was made very thin and lined with burlap, to combine strength and durability with the minimum of weight. After the plaster was dry, a coat of shellac was given to make it water-proof, the skin was adjusted, and the seams were neatly sewed up. Last of all, the eyes, nose and mouth were modeled,—the

COMPLETED PLASTER CAST OF LION
Ready for application of skin

most difficult and interesting part of the work, for the delicate lines require the utmost skill and closest study for successful reproduction, and the modeling here determines the whole expression of the face and the success or failure of the mount.

This, in brief, is the method which was employed in preparing the Carnegie lion for exhibition. In mounting the animal the subject has been treated from the artist's standpoint, and the effort is successful in getting away entirely from the mechanical side of taxidermy. The attitude chosen is rather unusual. The animal is represented as being in a position of rest, which gives an excellent opportunity for displaying the general anatomy which has been so carefully worked out by the sculptor.

THE NAOSAURUS, OR "SHIP-LIZARD."

NE of the most ancient as well as most grotesque of fossil reptiles is the Naosaurus, a skeleton of which has recently been placed on exhibition in the Dinosaur Hall. The animal was about eight feet long, a heavy-bodied, short-tailed carnivorous reptile with an enormous bony fin upon its back. The fin is composed

"HANNIBAL"

From the modeled mount in the Museum

of the spines of the vertebræ greatly elongated, and each spine bears a series of little cross bars, the arrangement suggesting the masts and yards of a square-rigged ship, whence the name of Nao-saurus or "Ship-Lizard." This remarkable specimen is a part of the Cope Collection of Fossil Reptiles which was presented to the Museum by President Jesup a few years ago. The bones were collected in the Permian beds of the Wichita river region, Texas, by Charles H. Sternberg.

The spines of Naosaurus spread out like the sticks of a fan and during life were probably connected by tough, horny skin, though not covered with flesh, for without some such connecting tissue the spines might easily be wrenched out of place, dislocating the backbone, since the fin is an extension of the vertebræ, unlike the fins of fishes, which are independent of the backbone, or the crest of the Iguana, which is simply an outgrowth from the skin. Although the large sharp teeth are well adapted to seizing and tearing the animal's prey, they are curiously ill-fitting, and apparently the jaws could not be tightly closed. The under side of the body was covered with bony scales.

The use of the great back fin has not yet been satisfactorily explained. It may have served partly to protect the backbone, always the most vulnerable part in such animals, but more probably it was chiefly ornamental. Suggestions that it served to conceal the animal by resembling, to the untutored eye of its prey, the reeds and rushes among which it lurked, or as a sail to enable it to traverse the waters of the Permian rivers and lakes, need not be taken very seriously.

Although clumsy and awkward looking in comparison with the more highly developed carnivorous reptiles and mammals of later periods, the Naosaurus was the most active and powerful predaceous animal of its time. A suggestion of its fighting habits is conveyed in the injury to one of the spines in this skeleton. This was broken and displaced during life, probably in some affray, and afterwards united by a growth of false bone. Several other specimens in the collection bear marks of injuries received during life.

With this skeleton the Department of Fossil Vertebrates enters upon the illustration of the fauna of the Age of Amphibians, which preceded the Age of Reptiles as that preceded the Age of Mammals in the history of the earth. In the Hall of Fossil Mammals may be seen the rise and development of the various races of quadrupeds which to-day inhabit the earth; while the Dinosaurs, in the Dinosaur Hall, and the Marine

THE SKELETON OF NAOSAURUS

Cope Collection

A great, flesh-eating lizard, eight feet long, from the Permian beds of Texas.
Collected by Charles H. Sternberg. Mounted at the American Museum in 1906
by A. Hermann.

MODEL OF NAOSAURUS

Executed under the direction of Professor H. F. Osborn by Charles R. Knight
in 1907.

Reptiles, in the corridor, belong to an earlier period during which Reptiles were the dominant animals of the world, and the Naosaurus and its contemporaries of the Permian Period are of that still more ancient time when Amphibians, related to the efts and salamanders of the present day, were the dominant animals and the reptile race was in its infancy. The splendid series of Permian fossils contained in the Cope collection, together with valuable collections more recently made for the Museum, will make a remarkable exhibit of these gigantic amphibians and primitive reptiles, which have heretofore been imperfectly known.

The extreme remoteness of this period may be judged from the estimate that the Naosaurus lived twelve million years ago, or twice as long ago as the Brontosaurus, six times as old as the Four-Toed Horse and two hundred times as old as the Mammoth and the Mastodon or the oldest traces of fossil Man.

The scientific description of the Naosaurus skeleton by Professor Osborn will shortly appear in the Bulletin of the Museum.

A NEW ESKIMO EXHIBIT.

HROUGH the great amount of excellent material brought in from the many expeditions to the Far North made by Commander Robert E. Peary, U. S. N., and the extensive whaling cruises of Captain George Comer the American Museum stands preëminent among all institutions along the lines of ethnological research amid Arctic peoples. The completeness of the material and data thus assembled has enabled the Museum to install a series of groups and cases which illustrate vividly the home and village life of the Central Eskimo, together with their utensils, implements and weapons and the methods of using them.

A large free space has been formed at the north end of the North Hall on the ground floor of the building near the entrance to the auditorium by removing two of the tall pier cases and substituting lower cases which are better adapted to the display of the material used. One result of the change is that from any point within the area the visitor may obtain a clear general idea of the whole exhibit. Resting places for visitors have been provided in the shape of two skin-covered sledges which were among the number used by Mr. Peary in his arctic work.

An imagination that is vivid enough to eliminate the comforts of the exhibition hall would enable a person, sitting upon **one** of these sledges to think himself among the interesting inhabitants of the land of cold

INTERIOR OF ESKIMO IGLOO
From group in Hall No. 108

and snow,— the land of the midnight sun in summer and of darkness in winter, save for the brilliant moonlight and the aurora borealis.

A general glance at the exhibit leads to an appreciation of the bleak characteristics of the land of the Eskimo. Almost no wood is to be seen, most of the objects having been made from some part of an animal, and bone, horn, tusk and skin have been ingeniously made to serve every purpose. The seal-oil lamp has been devised for giving light and heat. Methods of manufacture are illustrated by life-sized figures of men

ESKIMO WOMAN FISHING THROUGH THE ICE
From group in Hall No. 108

and women making or mending harpoons, harness, sledges and gar-ments. The practically complete absence of metal from the region leads to the employment of thongs in joining bits of wood or bone for making sleds, boxes and boats. The stray bits of precious iron or steel obtained by barter are used only for weapons or tools.

Vegetation in Eskimo lands is extremely scanty and is almost exclusively confined to mosses, rushes or, in some localities, a few low shrubs, none of which are suitable for the food of human beings. The environment, therefore, drives the Eskimo to hunting and fishing as a means of livelihood, and the center of the space is occupied by a small modeled group representing a man in the act of harpooning a seal, while a woman crouching by his side is a vitally interested spectator. The providing of food being the chief problem of existence in these far northern regions, two life-sized groups have been installed at the north side of the exhibit which typify two aspects of the universal occupation. One of these groups represents the interior of an igloo or snow house, where a woman is cooking by means of a seal oil lamp, while a child is creeping about the floor; the other group shows a woman fishing through a hole in the ice, under the lee of a wall of snow blocks which protects her from the cold, biting winds.

In spite of their adverse environment the Eskimo have developed a love for art, as is shown in the case devoted to carvings and engravings in ivory and bone. The walls and case fronts of the alcove are decorated with skulls, tusks and horns of the walrus, narwhal and wild reindeer, while the figure of a man in hunting costume in a kayak has been placed on top of one of the cases. An additional pleasing feature of the new installation is the arrangement of concealed electric lamps within the cases, by means of which a diffused but ample light is thrown on the specimens.

MUSEUM NEWS NOTES.

The annual meeting of the Board of Trustees of the Museum was held at the Metropolitan Club, Monday evening, February 11. The officers of 1906 were re-elected, namely:

President, MORRIS K. JESUP,
First Vice-President, J. PIERPONT MORGAN,
Second Vice-President, HENRY FAIRFIELD OSBORN,
Treasurer, CHARLES LANIER,
Secretary, J. HAMPDEN ROBB,
Director, HERMON C. BUMPUS.

In addition to the routine business of the meeting, a vote of thanks was passed to the members of the Peary Arctic Club for their generosity in presenting to the American Museum of Natural History the valuable collections made by Commander Robert E. Peary, U. S. N., on his recent expeditions to the Arctic under the auspices of the Club, and Miss Maria R. Audubon and Miss Florence Audubon were elected Life Members of the Museum in recognition of their gift of valuable sketches, drawings, plates and personal trophies of the ornithologist, John James Audubon.

A CABLEGRAM from Professor Osborn announces the auspicious starting from Cairo, on January 30, of his expedition into the Desert of Fayoum. He goes with valuable coöperation on the part of the Egyptian government and has every prospect of achieving important scientific results. Professor Osborn and two of his assistants in the Department of Vertebrate Palæontology left New York on January 5, as related in the February JOURNAL, to explore certain portions of the Fayoum desert for fossil mammals needed to fill gaps in the series illustrating several lines of evolution.

THE material brought back by Commander Robert E. Peary, U. S. N., was removed from the ship "Roosevelt" to the Museum during the latter part of January. This material, which comes to the Museum as the gift of the Peary Arctic Club, adds a large number of particularly desirable specimens to the collections from the Far North. Magnificent skulls and skeletons of walrus, narwhals, seals and musk oxen, an entire herd of pure white reindeer (a new species which has been named *Rangifer pearyi* by Professor Allen), clothing and implements of household use, hunting and fishing and sledges are among the items of this collection. The most interesting single piece from the popular point of view is perhaps the sledge with the help of which the new farthest north record was made and which Mr. Peary has christened the "Morris K. Jesup."

THE collection made by the Tjäder Expedition into British East Africa was received at the Museum during January. This material, which is wonderful in the extent, variety, size and perfect condition of its specimens, fulfills the announcements of success already made. A friend has made it possible for the Museum to acquire this collection, and it will receive full description in a subsequent issue of the JOURNAL.

A COLLECTION of fossil leaves from the Fort Union beds of Tertiary time has recently been received at the Museum. The specimens were gathered by Mr. Barnum Brown and his assistants in central Montana during the field season of 1906. The collection, which is said by Doctor F. H. Knowlton of the National Museum, a high authority on palæobotany, to be the finest he has ever seen from this deposit, comprises many remarkably fine examples of twenty-one known species belonging to fourteen genera, besides several new genera and species. Aside from its value from the point of view of the palæobotanist, the collection has great importance as a means of separating several geological horizons. The Fort Union beds contain beds of lignite, or brown coal, which are an important source of fuel. In some places this lignite has been ignited by spontaneous combustion or through some other natural agency, and the fires have burned for an unknown length of time, baking and fusing the clays above and below the coal until they look like brick, slag or volcanic scoriæ. Specimens of this material also were brought in by the expedition.

THE International Exhibition of Safety Devices which was open in the power room and adjacent corridor from January 29 to February 9 was the first affair of the kind ever held in this country, and it attracted a large number of visitors. There were about 300 entries of exhibits comprising all sorts of contrivances for the prevention of accidents and of injury from unavoidable accidents in street, house and factory. In connection with the exhibition lectures were delivered on February 1, 4 and 7 by Dr. W· H. Tolman upon "European Museums of Safety Devices and American Industrial Betterment," while on February 11 Dr. Josiah Strong lectured on "Safety for American Life and Labor."

THE legal holiday lecture of New Year's Day was given by Mr. Frank M. Chapman upon the topic "The Home Life of Flamingos" and was illustrated with some of the remarkable photographs from nature a portion of which were used in making up the flamingo group. The lecture on Washington's Birthday was by Mr. Harlan I. Smith upon "The Five American Nations: Conquerors of the Snow, Forest, Mist, Desert and Plain." The attendance at the four lectures given on the principal holidays of the winter was 2710, indicating the hold that this course has upon the public.

THE West Side Natural History Society held a special meeting in the West Assembly Hall of the Museum on the evening of February 7, when Mr. B. S. Bowdish of Demarest, New Jersey, gave an illustrated lecture upon "The Birds of Demarest, New Jersey."

During February, Mr. Frank M. Chapman delivered a series of eight lectures in the Lowell Institute Course at Boston. His topic was "The Distribution of Bird Life in North America."

THE annual meeting of the Board of Directors of the Audubon Society of the State of New York was held at the Museum Thursday afternoon, January 17.

LECTURE ANNOUNCEMENTS.

MEMBERS' COURSE.

THE second course of lectures for the season 1906–1907 to Members. of the American Museum of Natural History and their friends. The programme is as follows:

Thursdays at 8:15 P. M.

February 21.— FRANK M. CHAPMAN, "The Birds of Spring."

February 28.— RICHARD TJÄDER, "Hunting Big Game in British East. Africa."

March   7.— FREDERIC A. LUCAS, "Whales and Whaling."

March   14.— E. O. HOVEY, "Earthquakes; Their Causes and Effects."

March   21.— CLARK WISSLER, "Living with the Indians of the Plains."

PUPILS' COURSE.

	Mar.	Apr.	
Monday,	4	8.—	"Along the Historic Hudson." By G. H. Sherwood.
Wednesday,	6	10.—	"Life in the Far North." By H. I. Smith.
Friday,	8	12.—	"New York City in Colonial Days." By R. W. Miner.
Monday,	11	15.—	"The American Indians of today." By G. H. Pepper.
Wednesday,	13	17.—	"Commercial Centers of Europe." By E. O. Hovey.
Friday,	15	19.—	"Farming and Ranching in the United States." By G. H. Sherwood.

	Mar.	Apr.

Monday,   18   22.—"Travels in South America." By Barnum Brown.

Wednesday, 20   24.—"Natural Wonders of our Country." By R. W. Miner.

Friday,   Apr. 5   26.—"The Products of Our Mines." By E. O. Hovey.

This course of lectures is open to public school children accompanied by their teachers and to the children of Members of the Museum on the presentation of membership tickets. Particulars of the course may be learned by addressing the Director of the Museum.

PEOPLE'S COURSE.

Given in coöperation with the City Department of Education.

Tuesdays at 8 P. M.

A course of five lectures on the "Far Eastern Question" by Mr. Elwood G. Tewksbury, American Board of Missions, New York City.

March   5.— "The White Peril."

March 12.— "The Siege of Peking."

March 19.— "The Yellow Peril."

March 26 — "The New Far East."

April   2.— "Asiatic-American Reciprocity."

Saturdays at 8 P. M.

A course of nine lectures on "Electricity and Electrical Energy" by Professor John S. McKay of Packer Collegiate Institution, Brooklyn.

March   2.— "Relation of Electricity to Matter,— the Electron Theory."

March   9.— "Relation of Electricity to Energy. An Electric Charge and an Electric Current."

March 16.— "Electric Currents, or Electricity in Motion."

March 23.— "Thermal Relations of Electric Currents."

March 30.— "Chemical Relations of Electric Currents."

MEETINGS OF SOCIETIES.

The meetings of the New York Academy of Sciences and Affiliated Societies will be held at the Museum during March as usual. The programmes are issued in the weekly "Bulletin" of the New York Academy of Sciences and sent to the members of the several societies. Members of the Museum on making request of the Director will be provided with these circulars as they are published.

WARD'S GREAT B.UE HERON

Habitat group representing a scene in central Florida near the Indian River.

The American Museum Journal

VOL. VII   APRIL, 1907   No. 4

URING the past month the Museum has issued a new Guide Leaflet. This pertains to the extensive collections in the Peruvian Hall and is descriptive of the strange looking "mummy bundles" which come from Peru and their contents and the objects which have been found associated with them in the graves. The author is Mr. Charles W. Mead, of the Department of Ethnology, who has devoted several years to the study of Peruvian archæology and has made many interesting discoveries. The importance of the mummy bundles and the contents of the graves of the ancient Peruvians is enhanced by the fact that the people had no written language and that these objects are almost the only data that we have for studying the manners and customs of the interesting people that the Spaniards found inhabiting the Pacific slopes of the central Andes at the time of the Conquest. The Leaflet, which is No. 24 of the Museum series, will be sent free, upon request, to any Member of the Museum. Others may obtain the pamphlet at the entrance to the building or from the Librarian on payment of 10 cents.

HABITAT GROUPS OF BIRDS.

WITH this number of the JOURNAL we present illustrations of two of the new bird groups which have been mounted in the Hall of North American Ornithology (Gallery floor, north wing). These are part of a series which is being prepared with funds provided by the North American Ornithology Fund representing the birds of the continent in their natural surroundings, or "habitats."

One of these groups represents a family of Ward's Great Blue Heron in the swamps of central Florida near the Indian River. This bird, once abundant, has been almost exterminated by plume hunters.

The other group illustrated in this number of the JOURNAL is that of the Prairie Hen and represents several couples during the mating season, when the male goes through his peculiar antics of "drumming"

THE PRAIRIE HEN

Habitat group showing a scene in the Sand Hill Region of western Nebraska

and dancing. The material for this group was collected by Mr. Chapman in the Sand Hill region of western Nebraska near the town of Halsey last spring. Mr. Chapman considered himself very fortunate in witnessing this performance of the Prairie Hen at close range, thus securing the data from which the group has been constructed.

The backgrounds have been painted by Mr. Bruce Horsfall, who was sent by the Museum to make the requisite studies in the field and the accessories have been prepared at the Museum under the direction of Mr. J. D. Figgins.

THE MUSEUM'S NEW WHALES.

HE Museum is fortunate in having secured the skeleton of two Atlantic Right Whales (*Balœna biscayensis*) which were captured off the south shore of the eastern end of Long Island on Washington's birthday. Such an opportunity has offered itself but seldom before in a generation, and the animals too are becoming very scarce. Furthermore, when whales are taken on a cruise the skeleton is rarely preserved, particularly with all the small bones intact.

Early in the morning of February 22 three whales, one small and two large ones, were sighted about five miles from shore by the fishermen of Amagansett, L. I., and immediately a boat load of hardy men under the leadership of Captain Josh Edwards, an old time whaler, set out in pursuit, for the carcass of a large Right Whale means thousands of dollars to its captors. Meanwhile another boat load of fishermen from Wainscott, ten miles west of Amagansett, joined in the chase, which became an exciting adventure for the two small boats so far from land in mid-winter. Fortunately the weather was clear, though cold.

The Amagansett crew made fast to the cow whale and succeeded in killing her in the open ocean, but they had a hard task in towing the body to the beach. The bull made good his escape, but the Wainscott crew harpooned the calf and then had an easy time, for the wounded animal, also a female, headed straight for shore and landed high and dry on the beach.

The next morning the news of the capture of the whales was received at the Museum and two men from the Department of Preparation and

THE AMAGANSETT RIGHT WHALE
Stripping off the blubber

THE AMAGANSETT RIGHT WHALE
Lower jawbone, stripped of flesh but still attached to the head.

THE AMAGANSETT RIGHT WHALE
The upper jaw, showing whalebone in place

THE WAINSCOTT WHALE
Right flipper, edgewise view

Installation were sent by the first train with instructions to "get the specimens." Their experiences are best told in their own words.

"We reached the whales Saturday evening, and after bargaining for both skeletons and the whalebone of the larger specimen we stopped the work of stripping off the blubber, until we could make our measurements and get full data for the construction of life-size models. The big cow measured 53 feet from tip of nose to notch of tail, which is equal to the maximum size for this species as noted by F. S. True in his book on "The Right Whales of the Western North Atlantic." The whalebone of this individual is fully seven feet long, also a record size, and is unusually perfect. The calf measured about 40 feet in length and had whalebone three feet long in the longest part.

"The following day the whalers finished removing the blubber, and then we set to work cutting out the skeleton. This was a large undertaking since we were obliged to remove the flesh in rather small pieces in order not to lose any of the bones, and our labor was rendered more difficult and trying by the waves that broke over us most of the time while we were at work, and froze in picturesque icicles that we could not appreciate at the time.

"Wednesday we had to face a new difficulty, for the surf became heavy and began to bury the remaining bones of the Amagansett whale in the sand, whence it would have been impossible to recover them. We waited anxiously for low tide Thursday and then hastily constructed a rude cofferdam using ribs for piles and whale flesh for filling. This contrivance, with one man actively bailing water and another vigorously shoveling sand, enabled the rest of our force to secure the last bones of the great beast, after two hours of the hardest work imaginable.

"The Wainscott whale, being smaller and higher on the beach, had already been secured and nothing remained to do but to clean the bones thoroughly and ship them to the Museum, which finished a week of hard but satisfactory work."

In spite of its commercial value, the whalebone of the Amagansett specimen, weighing some 1700 pounds, was purchased by the Museum and will be mounted in proper position in the skeleton or the model. The whole series of whale material now at the Museum will, when mounted, make an exhibit the equal of which in its line is not yet to be found in this country. The whalebone of the Wainscott whale was not secured by the Museum.

THE EGYPTIAN EXPEDITION.

NCOURAGING news comes from the expedition into the Desert of Fayoum for vertebrate fossils. Professor Osborn writes, under date of February 11, that help from Lord Cromer and Director H. G. Lyons of the Geological Survey has supplied the American Museum party with full equipment of tents, tanks and other supplies needed for life in the desert. He says, in effect: "In five days instead of the ten days estimated beforehand we were ready, and I despatched Daoud Mahommet, who had been out every year with Beadnell and Andrews of the British Museum, around by rail to Tamia on the western edge of Fayoum with instructions to camp near the most easterly of the bone pits, which is about forty miles from the railroad. We left the Gizeh pyramids, twelve miles from Cairo, on Thursday morning, January 31, and that evening camped near the Sakhara pyramids, the tombs of ancient Memphis. Mr. H. T. Farrar, who had been detailed by Doctor Lyons to accompany us, joined our party here; so we were eight tents and twenty-one camels strong,— quite a big caravan and most picturesque.

"Friday we went beyond the Dashur pyramid, and the following night we camped at Lish't, where the Metropolitan Museum excavations are in progress under the direction of Doctor Lythgow. Sunday we traversed the desert and reached Tamia on the edge of the Fayoum oasis, and Monday night we made our first desert waterless camp. Tuesday, February 5, just a month from leaving New York, we reached our main desert camp and found that the men who had been sent around by rail had arrived two days before. We have secured seventeen diggers and eight camels for the transportation of water and supplies, and our camp of four tents under the charge of Mr. Granger and Mr. Olsen is between the two easterly bone pits.

"The country has been thoroughly prospected on the surface, but careful and extensive quarrying with the thorough methods which we have used so successfully in our own West is certain to produce good results. We have the best trained and the largest force of Egyptian workmen which has ever been at this locality, and our first search has produced several apparently new members of the smaller fauna, together with excellent jaws and isolated teeth and bones of the larger forms."

THE RESULTS OF THE TJÄDER EXPEDITION.

HROUGH the Tjäder expedition to British East Africa, mention of which has been made from time to time in the pages of the JOURNAL and the narrative of which has become familar to the Members through the lecture by Mr. Tjäder on February 28, the Museum has come into possession of a rich collection of zoölogical material comprising about 450 specimens among which are representatives of the principal types of the mammalian and bird fauna of that part of the continent.

Our Members will recall that this expedition left New York March 1, 1906, and proceeded to Mombasa on the east coast of Africa, where

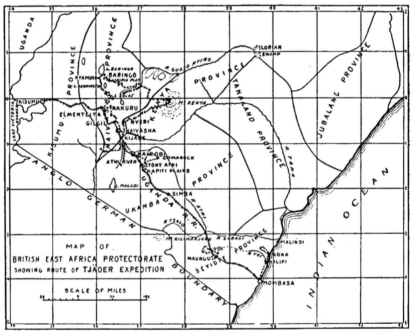

MAP OF PART OF BRITISH EAST AFRICA

Dotted line and railroad show route of Tjader Expedition

the Uganda railway was taken for the journey into the interior. Side trips were made from several points on the railway, and at the Athi river near Nairobi the party, consisting of Messrs. Tjäder and Lang and their assistants, began camping. Hunting was prosecuted for

H. Lang, Photó.

PART OF TJÄDER EXPEDITION CROSSING A RIVER

Near Nyamusi (Baringo).

ELEPHANT TAKEN NEAR NAKURU

R. Tjader, Photo.

The animal was 22 feet, 8 inches long, and his shoulder height was 10 feet, 4 inches. The tusks are 6 feet, 4 inches long and weigh 160 pounds.

59

H. Lang, Photo.

GIRAFFE SEVENTEEN FEET TALL

Obtained near Maungu. Photograph made after the animal had been shot, but before it fell.

H. Lang, Photo.

RHINOCEROS, 10 FEET 8 INCHES LONG, 5 FEET HIGH

Secured in the Valley of Solai, north of Nakuru

about four and a half months over a route which is indicated on the accompanying outline map of British East Africa.

A magnificent old bull Giraffe which stood with his head seventeen feet above the ground when alive, was secured only an hour's march from Maungu station on the line of the Uganda railroad. Another and smaller one shot near Comarock, Athiplains, 200 miles from Maungu, is interesting as showing the local range of the animal. A fine bull Rhinoceros was obtained in the Solai Valley, north of Nakuru and an unusually large Elephant, which goes as a trophy to Mr. Samuel Thorne, through whose generosity the expedition was made possible. The series of larger animals is completed by a cow Buffalo. The Buffalo are rather scarce, and they are so well protected that only the scientific collector can procure permission to shoot any. It is fortunate that the complete skeletons of the animals intended for mounting have been obtained and that the hides were taken off entire and not in several sections according to the usual practice. The weight of such hides is surprising, twelve to sixteen porters being needed to carry the fresh skin of an animal like the big Giraffe or the Rhinoceros, which weighed between 800 and 900 pounds. Some portions of the skin were two inches thick and it was necessary to pare as much as possible from the inside to prevent deterioration and permit subsequent manipulation for mounting. When ready for shipment the Giraffe skin weighed 120 pounds and that of the Rhinoceros 85 pounds.

Eight Zebras from different parts of the region and fifteen different species of Antelope are in the collection. The Antelope are particularly interesting and include excellent specimens of the Eland, the White-bearded Gnu, the Oryx, two kinds of Hartbeest, the Waterbuck, Reedbuck and Bushbuck, and several of the tiniest forms, such as *Cephalophus*, *Neotragus* and *Nanotragus*, of which the last is but eighteen inches long when full grown. The Waterbuck will make a particularly effective group, on account of the naturally proud aspect of the animals in life.

Mention should be made too of the specimens of Warthog, Spotted Hyæna, Jackal, Aard Wolf and Cerval Cat, while the Colobus Monkey noted for its beautiful fur, and other species of the quadrumana are represented by several specimens each. Taken together the trophies of this expedition supply exceptionally fine material of species which have hitherto been wholly lacking in the Museum collections.

Much interesting ethnological material was brought back by Mr. Tjäder. It consists of charms, ornaments, weapons, cooking utensils and other articles from several tribes. Long battle spears, ugly arrows bearing rows of jagged points and slender clubs with egg-shaped stone heads are among the things that show how the tribesmen contend with their enemies or attack wild animals.

The photographs illustrating this brief note were taken by Mr. Herbert Lang, the museum taxidermist who accompanied the expedition, and are used through the courtesy of Mr. Tjäder.

MUSEUM NEWS NOTES.

 EMBERS, and especially those having children, when visiting the Museum should be sure to avail themselves of the services of the instructor, Mrs. Agnes L. Roesler, who has recently been appointed by the Museum for the purpose of assisting Members and their guests. On coming to the Museum, Members may call for Mrs. Roesler, and she will accompany them through the exhibition halls and into the laboratories, where they may see how the artificial flowers, glass models, ethnological and other groups, fossil animals and other specimens are prepared for exhibition. The children of Members may be left with the instructor, who will be pleased to spend an hour or two with them in the exhibition halls, entirely at the convenience of their parents.

AMONG prominent foreign men of science who have been at the Museum during the past few months may be mentioned, Canon ARMOUR, of Worcester, England, Dr. T. TSCHERNYSCHEW, Director of the Imperial Geological Survey of Russia, Dr. E. TIETZE, Director of the Imperial Geological Survey of Austria, Prof. A. C. HADDON of Cambridge University, England, Prof. A. OSANN, Freiburg, Dr. T. IKI, of the Imperial Geological Survey of Japan, Prof. BRUNO WEIGAND, Strassburg, Prof. T. W. E. DAVID, University of Sydney, New South Wales, Dr. A. PLAGEMANN, of Hamburg, Profs. HUGO ERDMANN, E. PHILLIPPI, G. W. VON ZAHN and Dr. HECKMANN, of Berlin, Germany, Dr. H. SJÖGREN, Royal Natural History Museum, Stockholm, Dr. V. HACKMANN and Dr. W. WAHL, of Helsingfors, Dr. V. SABITINI, Rome, Italy, Mr. C. RAMALINGA REDDY, St. Johns College, Cambridge,

Eng., Dr. B. VON INKEY, of Tarótháza, Hungary, and Prof. D. RANDALL-MacIVER, of Oxford University, England.

MR. FRANK M. CHAPMAN left New York on March 8 in quest of certain birds whose nesting habits are inadequately represented among our groups. He will visit Florida, the Bahamas and Louisiana before returning to the Museum.

COMPLETE sets of the following publications have been added to the Library since January 1, 1907:
Atti della Società Romana di Antropologia in 12 volumes;
Annali del Museo Civico di Storia Naturale di Genova in 40 volumes;
Journal of the Maine Ornithological Society in 8 volumes;
Journal of the Anthropological Institute of Great Britain and Ireland in 32 volumes;
Froriep's Notizen in 85 volumes;
Isis von Oken in 41 volumes;
Recueil Zoologique Suisse in 5 volumes;
Revue Suisse de Zoologie in 10 volumes;
Bulletin de la Société Malacologique de France in 7 volumes;
Monthly Microscopical Journal in 18 volumes;
Archiv für Mineralogie, Geognosie, Bergbau und Hüttenkunde in 26 volumes.

A special view of the skeletons of the Amagansett and Wainscott Whales, the securing of which is described in this number of the JOURNAL, was made in the west court of the building March 17, the day after their arrival at the Museum. The popular interest in the great animals, the larger of which exceeds in size any known dinosaur, was shown by the crowds of people that came to see the skeletons.

————·————

LECTURE ANNOUNCEMENTS.

PUPILS' COURSE.

Mondays, Wednesdays and Fridays at 4 o'clock.
Friday, April 5 and 26.— "The Products of Our Mines." By E. O. HOVEY.
Monday, April 8.— "Along the Historic Hudson." By G. H. SHERWOOD.
Wednesday, April 10.— "Life in the Far North." By H. I. SMITH.

Friday, April 12.— "New York City in Colonial Days.' By R. W. Miner.
Monday, April 15.— "The American Indians of To-day." By G. H. Pepper.
Wednesday, April 17.— "Commercial Centers of Europe." By E. O. Hovey.
Friday, April 19.— "Farming in the United States." By G. H. Sherwood.
Monday, April 22.— "Travels in South America." By Barnum Brown.
Wednesday, April 24.— "Natural Wonders of Our Country." By R. W. Miner.

PEOPLE'S COURSE.

Given in coöperation with the City Department of Education.
Tuesday at 8 P. M. Illustrated. By Mr. E. G. Tewksbury.

April   2.— "Asiatic-American Reciprocity."

A course of four lectures on "The Evolution of the Japanese Nation" by Dr. William E. Griffis of Ithaca, New York.

April   9.— "Ancient Non-Mongolian Japan to 700 A. D."
April 16 — "The Making of the Japanese Nation, 700–1200 A. D."
April 23.— "Mediæval and Feudal Japan, 1200–1868."
April 30.— "Modern Japan. The Restoration of the Mikado. Adoption of the Forces of the West. 1868–1907."

Saturdays at 8 P. M.

Conclusion of a course of nine lectures on "Electricity and Electrical Energy" by Professor John S. McKay of Brooklyn.

April   6.— "Relation of Electric Currents to Magnetism."
April 13.— "Relation of Magnetism to Electric Currents."
April 20.— "Direct Currents, Generators and Motors."
April 27.— "Alternating Currents and Alternating Current Machines.'

MEETINGS OF SOCIETIES.

On Monday evenings, The New York Academy of Sciences:
   First Mondays, Section of Geology and Mineralogy.
   Second Mondays, Section of Biology.
   Third Mondays, Section of Astronomy, Physics and Chemistry.
   Fourth Mondays, Section of Anthropology and Psychology.
On Tuesday evenings, as announced:
   The Linnæan Society, The New York Entomological Society and the Torrey Botanical Club.
On Wednesday evenings, as announced:
   The New York Mineralogical Club.

THE VICTORIA FALLS OF THE ZAMBEZI RIVER

The American Museum Journal

Vol. VII	MAY, 1907	No. 5

THE DOUGLAS AFRICAN COLLECTION.

HROUGH the generosity of Messrs. Percy R. Pyne, Cleveland H. Dodge and Arthur Curtiss James, the Museum has acquired a large ethnological collection which was made recently by Mr. Richard Douglas in south central Africa. This acquisition is of partieular importance, not only on account of the great amount of material received, but also because heretofore the Museum has had few and only isolated specimens from the Dark Continent.

Africa is the primitive home of the negro race, representatives of which have been more or less a factor in the Occidental civilized world since the early days of Egypt. Upon the royal tombs and temples of Karnak, Luxor and Thebes we find in color and relief triumphal and other processions in which appear now and again among the captives or the slaves the unmistakable facial features presented by the negro of today, showing that there has been practically no change for thousands of years. The permanence of these characteristics is surprising to those who believe man to have come into existence within the last eight or ten thousand years of the earth's history. In spite, however, of this conservatism in feature, hair and complexion, the black peoples of Africa present great variety of anatomical, linguistic and tribal differences, ranging from the illusive pigmy of the Congo forest to the tall, clean, light colored Zulu of the South.

Along the Upper Nile and westward along the borders of the Sahara there is a broad belt of dark-skinned peoples where the lighter Arabian blood of the northeast gradually shades into the black of the Congo and the South. The arts and culture too of the Mediterranean states that followed the Arabic intrusion were gradually overwhelmed by the great monotony of native African barbarism. Yet for centuries, possibly while the savages of the stone age were hacking each other to pieces in primeval Europe, the peoples of the Dark Continent were smelting and forging iron, cultivating their fields with iron hoes and rising against their enemies with iron spears and swords. Study of Africa proves that an "iron age" is not of itself to be regarded as a guarantee of an advanced order of civilization. The effect of the use of iron implements is but one of the many interesting problems arising from the study of the Dark Continent, all of which render an ethnological collection from any of her people a matter of great educational value.

During the past year Mr. Douglas visited Barotse and Bechuanaland. As may be seen from the map, these territories occupy the entire central portion of that part of Africa lying between the southern borders of the Congo Free State and the Orange river. This region is cut into

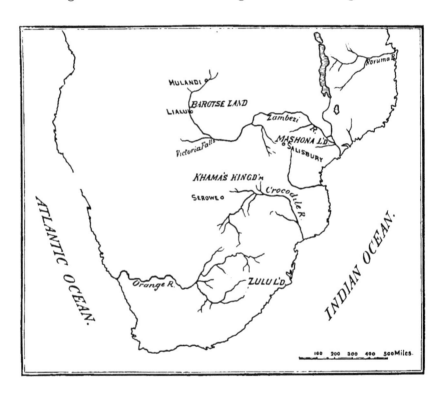

two parts by the Zambezi river, well known for its beautiful Victoria Falls. All the interior of South Africa has been for some centuries the home of a large division of the so-called "Bantu" peoples, the dominant negro race. It is generally agreed that the Bantu originated somewhere near the head-waters of the Nile. As they increased in numbers, they migrated southward and eastward, dominating the whole continent from the Sahara to the Cape of Good Hope. The Bantu horde which rolled out from the north into the valley of the Zambezi and into the

A BECHUANA VILLAGE

Kalahari Desert, is chiefly known under the name of Bechuana. As a matter of fact, however, this name belongs to a very small group of tribes that drove out the original inhabitants, who may have been the Bushmen, and took possession of their lands. The best known divisions of these Bantu intruders are the Bechuana, Zulu, Mashona, Barotse and Basuto.

Mr. Douglas writes in part regarding the expedition as follows:

"I reached Cape Town, S. A., about May 1, leaving directly for Bulawayo, South Rhodesia, arriving there May 5. After a short stay I left for Bechuanaland and arrived at Palapye Road at 10 o'clock in the evening of May 15. Early next morning I arranged with the Bechuana Trading Company for transportation to Serowe, King Khama's "stardt" or native village, located some 60 miles from Palapye Road, and set out, arriving on the third day at 1 P. M. The following day I called upon the king but learned that he was away, inspecting one of his cattle posts, and I did not meet him for three days. Our meeting was at six o'clock in the morning. The native custom is for all the Chief Headmen to meet at daybreak in the "kglotta" or Court Yard, to dispose of the native criminals, brought in the day before for trial. This meeting, as well as witnessing the disposition of criminals proved very interesting and also afforded me the very best opportunity of getting down to business with the king and all his people. After I had been introduced to the king by one of his grandsons and had made my purpose known, I was received very cordially and the king gave out word to all his chiefs to give me all the help I required. In this way I was enabled to make my collection without further trouble. The king gave me free transportation to the railroad for all my material, and after numbering and packing my collection I started upon my return journey.

"My next stop was at Salisbury, whence I went to the Mazoi District, Mount Darwin and the Inyanga Districts, all in Mashonaland. My collections here were not large. After packing and securing some twenty-five carriers, I left for Salisbury, camping with my carriers every night. In three weeks time I had covered nearly 500 miles on a bicycle, traveling over a very mountainous country with only Kaffir paths to follow. Upon my return to Salisbury I repacked my collections, paid my carriers and left immediately for Bulowayo.

"Upon the 20th of July I left Bulowayo for Barotse land, King Lewanika's Country, for whom I have acted as confidential Agent for the past four years. Upon my return to Africa, I had notified King

Lewanika that I intended visiting his country as soon as he could get boats down to Livingstone to take me up the river to Lialui, his capital. Upon my arrival at Victoria Falls, July 21, I immediately went to Old Livingstone to see Imasho, the king's Headman to see if the boats were waiting, but found that they were still on their way down, he having heard by a runner or messenger that the king was sending for me.

"That evening upon my return to the Falls Hotel I found a message telling me that the Induna in charge of the boats had arrived. I immediately made ready to leave for Old Livingstone. The next morning, July 28, I was up at break of day and found the king's carriers waiting for me. There were sixteen in all, eighteen including my interpreter and cook. Upon reaching Old Livingstone, which was late in the day, I camped for the night, making plans to leave at daybreak for the boats, which had to be left five miles farther up the river, on account of the dangerous rapids. We reached the boats at 10 A. M. and left immediately for Kazeungula the first important native village on our route, although there are many small kraals between. Only one white man trades with the natives in that village. After leaving Kazeungula and paddling two days we came to the great game country. Here I camped three or four days to secure food for the natives.

"August saw us again on our way up the river toward Lialui. After three days we reached Nilesia, which to my mind is the most beautiful spot on the whole river. Here the country is covered with thick bushes and abounds in lions. We could hear them roaring long before dark, and they kept up their noise all night. We had to keep big fires going to keep them away. Early next morning a Dutch transport rider came to my camp and asked me to assist him in hunting some lions which had killed five of his oxen. That night we took up the spoor and after following it for four miles, we came upon one lion, one lioness and two cubs. We got the lion and both cubs, but the lioness, although badly wounded, got away into the tall grass. We did not go after her, as it is a very dangerous undertaking to follow up a wounded lion, a thing that only inexperienced hunters will do, as there is only one chance in ten of getting away alive. After removing and caring for the skins, we proceeded on our journey, but since we had many rapids to cross, our progress was very slow. We reached more rapids next day about noon; here we had to take everything out of the boats and pull them overland, a distance of 300 yards. This took us until 4 o'clock, and we pitched

A GIRL GRINDING KAFFIR CORN

camp some four miles further up the River that night. On my way to this camp I collected many stone implements.

"We reached Lialui late in the afternoon of Thursday, August 28, having been one month in covering the 500 miles from Victoria Falls. During my stay at Lialui from August 28 to September 22, I was with the king daily and by his influence secured many fine specimens."

All the collections made by Mr. Douglas are now in the Museum. He has lived in South Africa for about twenty years and is not only familiar with the natives but also able to speak some of their languages. With the collection are numerous notes and other information of considerable ethnological value. The larger part of the collection is from Barotseland.

The Barotse kingdom extends from the vicinity of Victoria Falls on the Zambezi to the Congo Free State and eastward to the land of the Bashukulombwe, one of the recent conquests of the present king. It is now in control of King Lewanika under the protection of the British Government. While the Barotse are evidently a part of the great Bechuana group, their real relationship is not well known. They appear to be closely related to the Zulu, but they are generally looked upon by other Bechuana people as being the oldest original stem, from which the others sprang. However this may be, they are in many respects the strongest and most powerful group in Africa. Their history is exceedingly interesting and suggestive, one incident of which may be mentioned here. About 1835 the chief of Makalolo, a branch of the Basuto, extended his empire to the borders of the Barotse, made war upon them and soon brought them under his control. These foreign kings ruled until 1879, when the Barotse revolted and massacred all the Makalolo they could find. During their subjection, however, the Barotse had learned the Makalolo language, and that is still the official tongue. Thus we have the people of a kingdom speaking a foreign language to the exclusion of the native tongue, the whole change having taken place within a period of forty-four years. Since in other respects the history of the Barotse is similar to that of other native African tribes, the suggestion is that many times in the past peoples may have changed their language and customs in an equally rapid manner.

The greater part of the Bantu people in South and East Central Africa, support themselves by cattle raising, while the other Bantu tribes occupying the north and west portions of Africa support themselves by agriculture. The Barotse are situated on the dividing line between these two types of civilization, consequently we find what may be expected,— a people who are engaged in both agriculture and cattle raising. The raising of cattle, however, is almost entirely in the hands of the king and the various chiefs, practically no one being allowed to own cattle except the few head men. These cattle are the chief source for revenue for the native kingdom. The food of the common people is chiefly milk and the products of their gardens. While the men sometimes clear the fields, agriculture is almost entirely the work of the

A HOE FROM KHAMA'S KINGDOM, BECHUANALAND

The blade is about 2 feet long

women. The chief agricultural implement is the hoe. The hoe of Bechuanaland is a large leaf-shaped iron blade in a short wooden handle, while the hoes in Barotseland have small thin metal blades, similar to modern American hoes. The products of their fields are Kaffir corn (a kind of millet), Indian corn and yams. The villages are groups of circular, thatched huts, often clustered around the cattle kraal of a chief.

These people are skilled in the manufacture of pottery and wooden vessels. The wooden vessels are carved from tree trunks, hollowed out with the adze and finished with a peculiar hooked knife. The

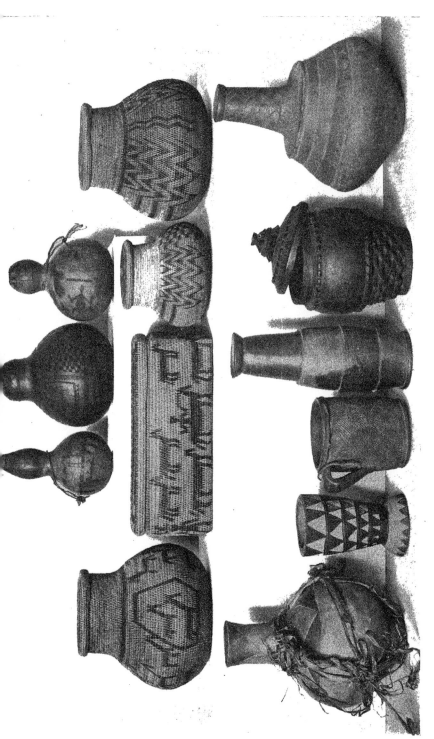

BASKETS, POTTERY, GOURDS AND WOODEN WARE

From Barotseland and Khama's Kingdom

surface of the vessel is burned with a hot iron and afterwards smeared with hot bee's wax, which is thoroughly rubbed into the wood, giving the surface a dead black finish. Perhaps the finest type of wooden vessel is the oval flat tureen with a lid. On the top of the lid is usually a carved elephant, hippopotamus or other large quadruped.

Pottery is of several forms, varying from large handsome jars to small drinking cups with handles. The most common type, however, is the water bottle. Two kinds of clay are used, which are mixed in certain proportions and modeled by a combination of the coil and beating processes. In the beginning, the bottom of the pot is fashioned in a shallow basketwork tray, which is turned with the left hand somewhat after the fashion of a potter's wheel. All of the pottery in the collection is red, but decorated with triangular designs in black or dark red. These designs are quite simple and consist usually of single or double rows of equilateral triangles. The same sort of decoration is applied to wooden ware, the triangles being produced by scraping away the previously blackened surface of the wood. The people from whom these collections came

A CARVED WOODEN STOOL

About 14 inches high

also manufacture black pottery of excellent quality, but this art is fast disappearing and no specimens could be obtained.

Wood carving is rather highly developed, the best types of which are to be seen in stools and sticks. While the common people usually sit upon the ground or upon mats, chiefs and other prominent people sit upon low wooden stools. These stools are usually cut from a single block of wood. One very common type is that in which the base and the top of the stool are joined by a human figure, supported behind by two or three upright posts. These figures always have the attitude of supporting the top of the stool upon the head. In some stools however, these figures are wanting, and the decoration consists of small geometrical designs arranged symmetrically on the posts. Wooden pillows are similar in form to the stools and the decorations are of the same type. Single pillows are usually used by the unmarried, while married persons use a double pillow, joined by a wooden chain cut from a single piece of wood. Aside from these, the collection contains a great many other carved objects such as a wooden figure, idols, combs for the hair and ear ornaments. Among the things deserving special mention, are the knob-sticks from Mulandi, which have a finish and execution far superior to anything else in the collection.

CARVED KNOB-STICKS FROM MULANDI

The collection is rich in basketry and matting. One of the most remarkable things about this basketry is the great variety of weave. In it we find wicker, checker, twill, close twine, open twine, twilled twine, three-ply twine, ti twine, one-rod coil, bifurcated coil, grass coil, open grass coil, coil without foundation and wrapped weave. Of these, the one-rod coil and open twine are the finest types. The decorations are in dead black, produced by steeping the material in marsh mud. It is of special interest to find the ti weave here, since this has heretofore been considered peculiar to the Pomo Indians of California. The collection contains one large storage basket similar to one shown on page 81. All such baskets are of the open grass coil type. The designs upon mats and baskets are triangular like those upon pots and wooden

MAKING A LARGE STORAGE BASKET

A Basuto man is in the basket

vessels, though occasionally the forms of animals and men are found upon baskets.

In south and central Africa the Barotse have great reputation as workers in iron, but their implements are crude. The smelting is done with a rude furnace, and the forging with rude bellows made of skin, stone anvils and in some cases with stone hammers too. Nevertheless with these crude tools the native blacksmiths turn out some excellent knives, daggers, axes, spears and swords. The collection contains a great variety of iron tools, spears and ear ornaments, illustrating quite completely the native iron industries.

A very conspicuous character in religious and ceremonial activities in all African tribes is the so-called witch doctor, who is in reality a priest. Such men have various outfits, consisting of charms, medicines and regalia, but in almost every case they have upon a string two slender pieces of ivory representing women and two hoofs of some ruminant representing men, together with two or more vertebræ of a monkey or other small mammal. The vertebræ are said to represent the spirit of the witch, as it is sometimes called, by whose help the priest accomplishes his work. This collection contains one complete witch doctor's outfit together with other medicine articles. The witch doctor is a powerful man in the community and performs various functions. Besides curing diseases, he discovers by magic processes the identity of criminals and traitors, directs all ceremonies and acts as chief councilor to the king or chief. The significance of his name is probably due to a widely spread belief in Africa that every death is the result of the magic power of some living person or witch. As a result of this the priest or witch doctor is called in to investigate every death, and as a rule he names some individual who is held responsible. It goes without saying, that witch doctors and chiefs take advantage of this custom to get rid of troublesome individuals. This is one of the many dark sides to the Dark Continent.

There are many other interesting groups of objects in this collection, among which may be mentioned native fibre, foods, costumes, weapons, pipes and musical instruments. The series of drums is particularly fine. The Museum now has a good beginning toward an African hall in which will be shown the original culture of the great Negro branch of the human family.

CLARK WISSLER.

DEPARTMENT OF MINERALOGY.

AMONG recent additions to the cabinet of minerals the new form of Beryl from near Spruce Pine, Mitchell Co., N. C., merits notice. This is an unusual tabular form of the mineral and was discovered by Mr. H. W. Williams. For some time it escaped proper identification on account of its peculiar crystalline form. The crystal consists of a broad basal plane and a hexagonal pyramid, the two united in thin plates inclosed in a coarse granitic matrix. Professor Moses of Columbia University has described this remarkable occurrence. The specimens are valued as crystallographic novelties.

A really superb specimen of Polybasite has been obtained through the Bruce Fund. The specimen was found in Sonora, Mexico, and brought to the Museum by Mr. A. B. Frenzel. It is a splendid group of lustrous, intersecting plates, the plates being tabular prisms with pyramidal edges.

The third notable addition is a unique and particularly beautiful specimen of crystallized Native Copper. It is a thicket of nail-like, elongated prismatic crystals, possibly tetrahexahedrons, with minutely dentate edges, of brilliant surface, and associated with thickly clustered individual crystals. This specimen came from Bisbee, Arizona, where it was found in a pocket with other similar specimens of an inferior quality. It is implanted on a limonitic base. The specimen is not large, but its effectiveness as a mineral development is remarkable.

A specimen of crystallized Andorite from Oruro, Brazil, also secured through the Bruce Fund, is astonishingly good. Large, heavy tables in this specimen replace the diminutive crystals usually associated with this interesting sulph-antimonide of lead and silver. Pink Beryls from Haddam, Native Lead (F. A. Canfield) from Sweden, Serpentine and the famous Asbestos (Chrysotile) from the Grand Canyon of the Colorado (F. F. Hunt), with a series of attractive Japanese specimens, obtained by exchange with Professor T. Wada of Tokio, Japan, should also be mentioned.

L. P. G.

MUSEUM NEWS NOTES.

¦ THE MUSEUM is now open free to the public on every day of the week. This important change in the policy of the institution has been made by President Jesup in order to extend its usefulness as widely as possible, it being felt that the reservation of two days in the week, as heretofore, for Members and students was depriving thousands of people of the privilege of seeing the collections, without compensating advantages to Members, while students are now amply provided for at all times in other ways.

PROFESSOR H. F. OSBORN returned March 31 from his trip to Egypt to organize the work which the Museum is carrying on there in the search for the remains of the ancestors of the Elephant and other mammals. Messrs. Granger and Olsen have remained in the desert of Fayoum to prosecute the excavations. Professor Osborn reports excellent initial success and bright prospects.

MISS ADELE M. FIELDE has presented to the Department of Ethnology a series of twenty-seven Chinese paintings representing various mythical and real scenes from Chinese life. These paintings were made by a native artist in 1888 at the suggestion of Miss Fielde and were used by her as illustrations in her books, "Chinese Night's Entertainments" and "Corners of Cathay."

THE Department of Ethnology has recently received from Mr. Edward J. Knapp a series of wooden masks from the Eskimo of Point Hope, Alaska. Among them are several interesting portraits, done with remarkable skill, and several ceremonial masks with markings representing the flukes of the whale.

DIRECTOR H. C. BUMPUS represented the Museum at the ceremonies connected with the dedication of the new buildings of the Carnegie Institute, in Pittsburgh, April 11–13.

DR. ALLEN represented the Museum at the spring meeting of the National Academy of Sciences in Washington April 16 to 18.

ELABORATE preparations have been made by the New York Academy of Sciences for the appropriate celebration, on May 23, of the two hundreth anniversary of the birth of the celebrated Swedish naturalist, Linnæus. The exercises will begin in the morning at the American Museum of Natural History with addresses and an exhibition of the animals, minerals and rocks first classified by Linnæus; will continue in the afternoon at the Botanical Garden and Zoölogical Park, with addresses and suitable exhibits of plants and animals and the dedication of the Bridge, and will be concluded in the evening with simultaneous exercises at the Museum of the Brooklyn Institute, Eastern Parkway, and at the New York Aquarium in Battery Park. The exercises at the Museum will include, at 11 o'clock, an address by Mr. Archer M. Huntington, President of the American Geographical Society, on "North American Geography at the Time of Linnæus" and one by Dr. Joel A. Allen, Curator of Mammalogy and Ornithology at the Museum, on "Linnæus and American Zoölogy," while Dr. E. O. Hovey, Secretary of the New York Academy of Sciences, will read letters concerning the anniversary from other societies.

A FINE collection of European Myriapods, comprising 238 species has recently been acquired by the Department of Invertebrate Zoölogy, They were collected by Dr. Carl W. Verhoeff of Dresden, Germany, and embrace specimens from Germany, Austro-Hungary, Greece, the Pyrenees, European Turkey, France, Switzerland, Portugal, Italy, and Norway, together with a few from Tunis (Africa). The peculiar animal forms comprised in the class *Myriapoda* are familiar to all under such names as centipedes, millepedes and "thousand-legged worms." Like the true worms their bodies are long, cylindrical or flattened, and they are divided into a varying number of ring-like segments. They differ from the worms, however, in possessing one or two pairs of jointed legs for each segment, while their jaws, antennæ and internal organs closely resemble those of insects. Standing thus as an intermediate or transitional link between these two groups, myriapods are of peculiar interest to biologists. The centipedes, which differ from the millepedes in having but one pair of legs for each segment instead of two, are carnivorous and kill the insects upon which they feed by their poisonous bite. The poison also serves as a protection against enemies. The millepedes on the other hand are vegetarian in their habits, and therefore harmless, though some species are obnoxious to farmers because of the damage they work to crops.

MR. FRANK M. CHAPMAN, Associate Curator of Mammology and Ornithology, spent a fortnight during April visiting several of the Bahama Islands for the purpose of collecting nests, eggs and young of certain birds for the habitat groups now being prepared at the Museum. The authorities of the Carnegie Laboratory at Dry Tortugas, Florida, placed at Mr. Chapman's disposal the yacht "Physalia" and Dr. A. G. Mayer, director of the laboratory, accompanied him on the trip.

MR. J. D. FIGGINS, of the Department of Preparation and Installation, left New York on April 6 for Key West, Florida, where he will join Mr. Chapman for additional field work in Florida. From Florida the expedition will go to Louisiana.

THE Department of Mammalogy has recently acquired by purchase a collection of mammals from China. The series includes 106 specimens, mostly of species the size of a Hare or larger, of which 43 are from the Island of Hainan and 63 from the interior of China, near the foot of the Taipashiang Mountains. The latter are all new to our collection, and the Hainan specimens do not duplicate the material previously received from that island.

ON March 29 a delegation of about forty teachers from Buffalo visited the Museum and spent considerable time under guidance in studying the work carried on here in connection with the schools of this city. The system of lectures to children, the circulating nature study collections and other educational work of the institution were explained and demonstrated to the visitors.

THE National Kindergarten Association opens an exhibition at the Museum on May 2 which will continue through the space of three weeks.

THE next number of the JOURNAL will be issued in October.

MEETINGS OF SOCIETIES.

MEETINGS of the New York Academy of Sciences and Affiliated Societies are held at the Museum according to the following schedule:

On Monday evenings, The New York Academy of Sciences:
First Mondays, Section of Geology and Mineralogy.
Second Mondays, Section of Biology.
Third Mondays, Section of Astronomy, Physics and Chemistry.
Fourth Mondays, Section of Anthropology and Psychology.

On Tuesday evenings, as announced:
The Linnæan Society, The New York Entomological Society and the Torrey Botanical Club.

On Wednesday evenings, as announced:
The New York Mineralogical Club.

The programme of meetings of the respective organizations is issued in the weekly "Bulletin" of the New York Academy of Sciences and sent to the members of the several societies. Members of the Museum on making request of the Director will be provided with these circulars as they are published.

The meetings will be held throughout May and will then be discontinued for the summer, beginning again October 7 with the business meeting and section of geology and mineralogy.

SPECIAL NOTICE TO MEMBERS.

A LECTURE WILL BE GIVEN BY

COMMANDER ROBERT E. PEARY, U. S. N.

AT THE MUSEUM ON

TUESDAY, MAY 14, 8.15 P. M.

REGARDING

"The Work of the Peary Arctic Club in 1905 - 1906 and the Plans for 1907 - 1908."

NOTE.—The Auditorium will be reserved for members of the Museum and their guests.
Special cards of admittance will be issued.

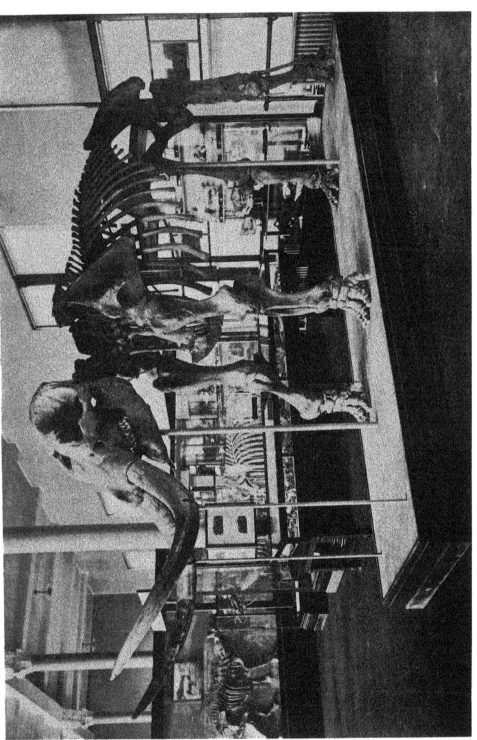

THE WARREN MASTODON. GIFT OF J. PIERPONT MORGAN, ESQ.

102

The American Museum Journal

| Vol. VII | OCTOBER, 1907 | No. 6 |

THE WARREN MASTODON.

N page 90 we present an illustration showing the Warren Mastodon, *Mastodon Americanus*, as installed in the Hall of Vertebrate Palæontology. This famous skeleton, the most complete which has been found, was discovered during the unusually dry summer of 1845 on the farm of Mr. N. Brewster in a small valley near Newburgh, N. Y. The bones were in an almost perfect state of preservation, and from the fact that they were buried in a layer of shell-marl, they were not black, like most mastodon bones, but brown, like those of a recent skeleton which has been much handled.

The bones were exhibited for three or four months during the same year in the city of New York and in several New England towns and were then purchased by John Collins Warren, M. D., who was a distinguished professor of anatomy in Harvard University from 1815 to 1847. In 1846 the skeleton was mounted, under Professor Warren's direction, by N. B. Shurtleff, in Boston, and exhibited to Sir Charles Lyell, Professor Jeffries Wyman, Professor Louis Agassiz and thousands of visitors. In January, 1849, it was remounted and placed with other collections in the fire-proof building on Chestnut Street, Boston, subsequently known as the Warren Museum, which was erected expressly for it. Here it remained till 1906, when it was acquired with the remainder of the Warren collection of fossils and presented to the American Museum by J. Pierpont Morgan, Esq., as was noted in the JOURNAL for April, 1906.

A year has been devoted to the work of renewing and remounting. The skeleton was taken apart and the dark varnish with which the bones had been covered was removed by the use of alcohol. Thus the original color of the time of discovery has been regained. The tusks were erroneously reported to Professor Warren as being more than 11 feet in length, and were so described and restored by him; but the original length has been exactly determined by skillfully piecing the fragments together as 8 feet 6 inches. Twenty-three inches of each tusk is inserted in the sockets, the projecting part measuring 6 feet 7 inches. The skeleton is so nearly complete that almost no restoration or replacement has proved necessary.

The following careful measurements will be of interest:

	Feet.	Meters.
Length, base of tusks to drop of tail . .	14 ft. 11 in.	4.55
Height to top of spines of back at the shoulders	9 ft. 2 in.	2.80
Tusks: Length of right tusk, on outside curve	8 ft. 6 in.	2.59
Length of tusk exposed . . .	6 ft. 8 in.	2.03
Thigh bones: Length of right . . .	3 ft. 5 in.	1.05
Length of left . . .	3 ft. 6½ in.	1.03
Pelvis, or innominate bones, width of . .	6 ft.	1.83

The Mastodon was the contemporary of the Mammoth in North America during Pleistocene or early and middle Quaternary time. Comparison of this specimen with the fine skeleton of the Mammoth, *Elephas columbi*, standing near shows the likenesses and the points of difference between the two animals. The Mastodon was generally longer, somewhat lower and more massive than the Mammoth. The most easily recognized difference lies in the teeth, those of the Mammoth showing low narrow transverse ridges, while those of the Mastodon show strong cusps.

A BLACKFOOT LODGE, OR TEPEE.

THE illustration on page 93 shows the lodge, or tepee, of a "medicine man" of the Otter clan of the Blackfoot Indians of Montana which has been installed in the Hall of North American Indians, No. 102 of the ground floor of the Museum. The lodge was obtained on a Museum expedition in the field season of 1903 by Dr. Clark Wissler, Curator of Ethnology, who is a regularly adopted member of the Blackfoot tribe. The decorations on the outside of the lodge represent the otter (the insignia of the family or clan), together with mountains (the triangular points) and stars (the white circles), while the black and red at the top are the signs of night and day.

The arrangement of the interior is such as to indicate the family life of the medicine man. The woman wears a typical Blackfoot costume and is engaged in performing ordinary home duties. Behind the

A B.ACKFOOT .ODGE, OR TEPEE

Group in North American Indian Hall, No. 102 of the Ground Floor

household fire is the family altar, which is only a patch of ashes where offerings of incense are sprinkled at certain times upon live coals from the fire. At the left of the altar may be seen the usual tobacco board and pipe, the sign of hospitality. At the right and left are the beds, which are made of and covered with buffalo hides. At the head of each bed is the back rest, suspended to the tripods of which are the "medicine" bags containing charms for use on ceremonial occasions. Among other articles in the lodge are household utensils, a man's saddle and parflèche bags for storing pemmican.

The background of the Otter lodge is formed by another Blackfoot lodge-cover, which was obtained by Dr. George Bird Grinnell. Both lodge-covers are made from cowskin which has been tanned and prepared in the usual Blackfoot manner, as illustrated and described in the exhibits on the opposite side of the hall.

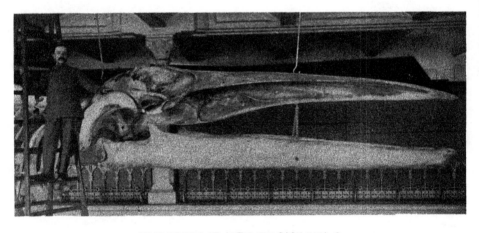

HEAD OF THE ATLANTIC FIN BACK WHALE
North Hall, No. 308 of the Third Floor

THE MUSEUM WHALES.

IMPORTANT additions have been made recently to the exhibition series of Cetaceans through means provided by George S. Bowdoin, Esq. Among these are the skeletons of three species of Whale which have been mounted in the East Mammal Hall of the gallery floor (Hall No. 306); a life-size model of the Atlantic Sulphur-bottom Whale-

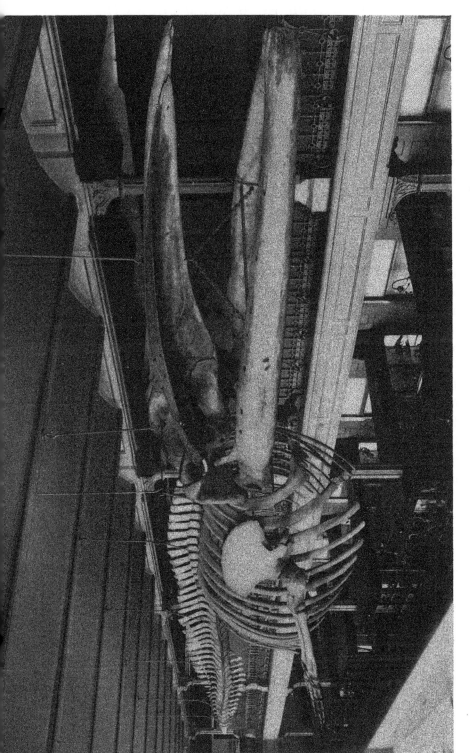

THE ATLANTIC FIN BACK WHALE, OR RORQUAL.   GIFT OF GEORGE S. BOWDOIN, ESQ.

The skeleton is $62\frac{1}{2}$ ft. long.   North Hall, No. 308 of the Third Floor

installed in the same hall, and a complete skeleton of an Atlantic Fin-back Whale, or Rorqual, which has been suspended from the ceiling in the North Hall (No. 308) of the third floor.

The model of the Sulphur-bottom Whale represents an animal 76 feet long in the act of swimming. It consists of papier maché upon a wire shell which has been built over an elaborate frame of structural iron. The Sulphur-bottom is the largest of marine mammals and, in fact, of all known animals either living or extinct, sometimes attaining a length of 95 feet, with a girth of 39 feet and an estimated weight of 147 tons. This whale occurs in the Atlantic as well as in the Pacific Ocean, but it has become very rare in recent years on account of relentless hunting. It receives its name from the color of the under surface. The skeletons in the East Mammal Hall are of a *Hyperoödon* or Bottle-nose Whale which was captured twenty years or more ago in the German Ocean, a *Globicephalus* or "Caá-ing Whale," as it is called by the Scotch, which was caught near the Faroe Islands, and a *Mesoplodon*, or Beaked Whale, which was taken near New Zealand in 1893.

The Fin-back Whale was captured off Provincetown, Mass., in April, 1896, and was about 63 feet long when alive. The maximum size for females of this species, which are larger than the males, is 70 feet. The Finback is still captured in considerable numbers off the coast of North Carolina and northward to Newfoundland. The whale is hunted by means of steamships and is killed with explosive harpoons. The commercial products obtained are whalebone of short length and coarse quality and oil, while the flesh and skeleton are used in making fertilizer.

AN EXHIBITION OF MUSEUM ART AND METHODS.

URING the month of May there was held in the East Mammal Hall an exhibition of drawings, paintings and models by the artists of the scientific staff of the Museum, showing the manner of preparing groups and figures for the public cases and illustrations for its scientific publications.

Among the features of the exhibition were studies in clay by James L. Clark for the mounting of the African Lion Hannibal and the group of Mountain Sheep and the Whales. Dr. B. E. Dahlgren showed a group of the Snapping Turtle and enlarged models of several minute forms of animal life. W. C. Orchard exhibited models of Indian heads of different tribes, colored to illustrate several styles of face painting used in ceremonies like the Ghost, Corn and Buffalo dances. The heads themselves were modeled by Caspar Mayer, who also exhibited several groups illustrating the Eskimo and the African Negro.

Charles R. Knight was well represented with sketches in water colors, oils and clay for some of the famous restorations of fossil mammals and reptiles which he has constructed under the direction of Professor Osborn. Bruce Horsfal exhibited field studies made for the backgrounds of several Habitat Groups, particularly those for the Prairie Hen, the Pelican, the Wild Turkey, the Anhinga and the birds of the desert. Albert E. Butler and Miss French contributed models in wax of flowers, fruit and foliage of North American trees from the series in course of preparation for the Jesup Collection of Woods and a charming little group showing the South American Flying Lizard in its home surroundings of orchids, butterflies and moths. Ignaz Matausch showed some enlarged water color drawings and wax models of the brightly colored, peculiar and little known insects called Leafhoppers.

The exhibition also included water color landscapes and minutely accurate water color reproductions of moths and butterflies by Mrs. E. L. Beutenmüller, careful pen drawings of ants by Miss M. E. Howe, pen and wash drawings of vertebrate fossils by Mrs. L. M. Sterling, Erwin S. Christman and B. Yoshihara and pen drawings of Indian relics by W. Baake.

NEW FEATURES OF THE EXHIBITION HALLS.

AMONG the recently installed features of the exhibition halls the following may be mentioned: Department of Vertebrate Palæontology,— Allosaurus group, Trachodon skeletons, Warren Mastodon skeleton, skeleton of Ichthyosaurus preserving impression and outline of body, additions to Horse Alcove, alcove labels; Department of Geology,— polished blocks of orbicular diorite and other rocks; Department of Mammalogy,— model of Sulphur-bottom Whale, skeletons of Atlantic Finback and other whales; Department of Ornithology,

— the Wild Turkey group, the group of Feeding Birds; Department of Ethnology,— Blackfoot lodge, Maori heads, Japanese reception room; Department of Invertebrate Zoölogy,— enlarged models (75 diameters) of the Malaria Mosquito, model of the North Atlantic Squid, alcove labels on glass.

A DIPLODOCUS FOR THE FRANKFURT MUSEUM.

HE illustrations on this page and the following are of the new Senckenberg Museum of Natural History which has just been finished at Frankfurt on the Main, Germany, and the interior court of the building. In this court as the place of honor in the museum has been installed a skeleton of the great fossil herbivorous reptile,

THE SKELETON OF DIPLODOCUS

Mounted in the covered court of the Senckenberg Museum

Diplodocus, a gift from Mr. Morris K. Jesup. This specimen, which is sixty-one feet long and twelve feet high, was taken from the famous Bone Cabin Quarry, near Medicine Bow, Wyoming, the place from

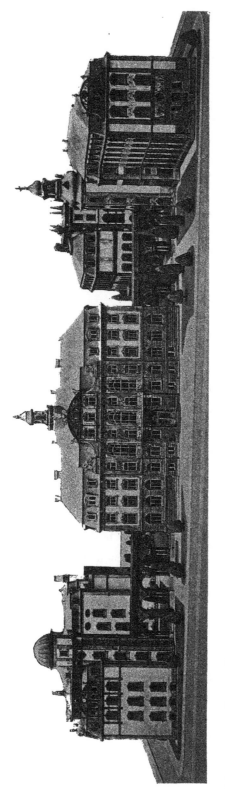

PHYSICS SOCIETY   SENCKENBERG NATURAL HISTORY MUSEUM   LIBRARY

THE NEW BUILDINGS OF THE SCIENTIFIC SOCIETIES, FRANKFURT ON THE MAIN, GERMANY

The Diplodocus given to the Museum by Mr. Jesup has been mounted in the court of the central structure.   See page 98

which the American Museum has secured its extensive and almost unique series of remains of Diplodocus, Brontosaurus and other gigantic reptiles which flourished in the shallow lakes and marshes that characterized the eastern portion of a part of the Rocky Mountain region in Jurassic time, some eight millions of years ago. This is the first skeleton of its kind to be sent to Europe, and the gift is made in the hope that it may be instrumental in bringing the museums of both continents into closer relationships and that it may foster the kindly feeling now existing between the German and American peoples.

The Senckenberg Museum will be formally dedicated October 13 with elaborate ceremonies under the patronage of the Emperor and Empress of Germany. The American Museum will be represented on the occasion by Director Bumpus.

THE ROBLEY COLLECTIONS OF MAORI HEADS.

THE Department of Ethnology is particularly fortunate in having secured last May the remarkable and practically unique collection of tattooed heads of ancient Maoris which Major General G. Robley of the British Army spent many years in assembling at infinite pains and great expense in New Zealand and from other authentic sources. These heads, thirty five in number, illustrate all the different styles of the art of tattooing as practised among the Maoris prior to the year 1831. At that time the British government forbade further tattooing, because the high value set on the heads by souvenir hunters led to the commission of many murders. A full description of the series of heads and of the outfit of ancient tattooing tools received therewith is reserved for later publication.

MUSEUM NEWS NOTES.

THE great meteorite known as Ahnighito which Commander R. E. Peary, U. S. N., secured in the summer of 1897 on the shores of Cape York in northern Greenland was transferred in August from the position which it has occupied for about two years under the arch at the entrance to the Museum to its permanent abiding place in the Foyer.

The task of moving this $36\frac{1}{2}$ ton mass of iron to its present position with all the resources of the city at command and with plenty of time for the work has made us realize more than ever before the bravery and skill shown by Mr. Peary in bringing the meteorite away from its Arctic home. A thrilling account of Mr. Peary's expedition for the Saviksue or Cape York meteorites may be found in his book "Northward over the great Ice," and a brief notice of the three irons, Ahnighito, the Woman and the Dog, comprising the group may be found in the AMERICAN MUSEUM JOURNAL for January, 1905.

THE Gem Collection has received as a gift from Mr. J. Pierpont Morgan, a boulder of jade (nephrite) from New Zealand weighing 7,196 pounds. This is the largest single mass of this material which is known to be in existence.

THE Department of Mineralogy received in August a valuable gift of Brazilian gems and gem material from Mr. J. F. Freire Murta of Arassuahy, Minas Geraes, Brazil. The series consists of cut gems and unworked fragments illustrating the valued colors of tourmaline and beryl occurring in the state of Minas Geraes.

THE Seventh International Zoölogical Congress, which held its scientific sessions in Boston August 19 to 24, was the guest of the American Museum on Tuesday, August 27. At eleven o'clock the officers of the Museum met the members and delegates of the Congress in the Foyer and conducted them through the exhibition halls, pointing out the particular zoölogical treasures. Among these the collections of the departments of Vertebrate Palæontology and Invertebrate Zoölogy attracted the most attention. At one o'clock the members of the Congress were the guests of President Jesup at a luncheon which was served in the corner hall opening out of the Laubat Hall of Mexican Archæology. The afternoon was spent in visiting the laboratories and work rooms of the Museum where the "congressists" were particularly interested in the work being done in glass, wax and other materials in the preparation and mounting of groups and individual specimens. During the evening a reception was given in the building by the Trustees of the Museum and the Council of the New York Academy of Sciences, when the foreign and out-of-town delegates had an opportunity of meeting New Yorkers who are interested in science non-professionally.

An attractive feature of the reception was the series of exhibitions of stereopticon views illustrating recent field work of the Museum and associated institutions in the Fayoum Desert, East Africa, the Bahamas and elsewhere.

THE Department of Mammalogy has recently obtained the skins and complete skeletons of two specimens, a male and a female, of the extremely rare *Solenodon paradoxus* which were collected by Mr. A. H. Verrill in the island of Haiti during the early part of this year. The Solenodont, called the Agouta in Haiti, is a small insect-eating animal, rarely more than twenty inches in total length, with a long naked nose and a long scaly tail and strong claws. Heretofore it has been known in museums by a single skin and skull which are in St. Petersburg, and even the Cuban Solenodont, though more common, is found in but few collections. Another important recent accession in this department is the skeleton and skin of an adult Sea Otter, *Latax lutris nereis*, which was captured in the latter part of last July near Point Lobos, California. The skin is five feet two inches long from tip of nose to tip of tail, but the animal may have been longer than this when alive, since the skin has been stretched sidewise. The Sea Otter ranged formerly from the Bering Sea southward along both coasts of the Pacific Ocean. On the east coast its range extended to northern Lower California, but the animal has become nearly extinct on American shores, and a hunter considers himself well repaid for a year's search by the taking of a single fine specimen.

THE Japanese Room, which attracted much attention in the Japanese government exhibit at the Louisiana Purchase Exposition at St. Louis in 1904, has been recently opened to the public in the Southwest Hall (Hall No. 201) of the second floor of the Museum. The room is richly decorated in silk, carved native woods and lacquer to illustrate the adaption of oriental materials and patterns to occidental uses. This exhibit has been presented to the Museum by the Nippon Yusen Kaisha through Baron Kaneko of Japan.

MR. FRANK CHAPMAN, Associate Curator of Mammology and Ornithology, accompanied by Mr. J. D. Figgins of the Department and Mr. Bruce Horsfal, the artist, visited the coast of South Carolina in May for the purpose of collecting material for the Egret group. Mr. Chapman made another expedition in June and July with the artist Mr. L. A. Fuertes to Saskatchewan for the wild water fowl of the Northwest and to the Canadian Rockies for Ptarmigan. The expeditions were eminently successful in procuring the skins, accessories, photographs and sketches needed for the groups, which form part of the series of Habitat Groups provided for by the North American Ornithology Fund.

MESSRS. WALTER GRANGER and George Olsen of the Department of Vertebrate Palæontology, returned July 4 from Egypt, where they had spent more than four months in active excavation and exploration in the Fayoum Desert. The objects of the expedition, which was under the immediate direction of Professor Osborn, were set forth in the AMERICAN MUSEUM JOURNAL for last February. The results were highy satisfactory, but a detailed notice of them is reserved for a later number of the JOURNAL, after the material shall have been received at the Museum.

LECTURE ANNOUNCEMENTS.

MEMBERS' COURSE.

THE first course of lectures for the season 1907–1908 to Members of the Museum and persons holding complimentary tickets given them by Members will be held in November and December. The lectures will be delivered on Thursday evenings at 8:15 o'clock and will be fully illustrated by stereopticon views. The programme will be announced this month in a special circular.

PUPILS' COURSE.

THE lectures to Public School children will be resumed in October. These lectures are open to the pupils of the public schools when accompanied by their teachers and to the children of Members of the Museum on the presentation of Membership tickets. Additional particulars of this course may be learned by addressing the Directors of the Museum.

PEOPLE'S COURSE.

Tuesday evenings at 8 o'clock.

A course of lectures illustrated with stereopticon views.

October 1.— DR. P. H. GOLDSMITH, "The Great Mexican Cornucopia."

October 8.— Mrs. Alice D. Le Plongeon, "The Famous Ruins of Yucatan."

October 15.— Professor William Libbey, "Cuba."

October 22.— Mr. Orrel A. Parker, "Porto Rico and Its People."

October 29.— Mr. George Donaldson, "The West Indies."

Saturday evenings at 8 o'clock.

A course of three lectures by Professor Samuel C. Schmucker illustrated with charts and specimens.

October 5.— "Crabs and Their Cousins."

October 12.— "Insect Changes."

October 19.— "A Family of Spinners (Spiders)."

October 26.— D. Everett Lyon, Ph. D., "The Life Story of the Honey Bee."

MEETINGS OF SOCIETIES.

Meetings of the New York Academy of Sciences and Affiliated Societies are held at the Museum from October to May, inclusive, as follows:

On Monday evenings, The New York Academy of Sciences:

First Mondays, Section of Geology and Mineralogy.

Second Mondays, Section of Biology.

Third Mondays, Section of Astronomy, Physics and Chemistry.

Fourth Mondays, Section of Anthropology and Psychology.

On Tuesday evenings, as announced:

The Linnæan Society, The New York Entomological Society and the Torrey Botanical Club.

On Wednesday evenings, as announced:

The New York Mineralogical Club.

On Friday evenings, as announced:

The New York Microscopical Society.

The programme of meetings of the respective organizations is published in the weekly "Bulletin" of the New York Academy of Sciences and sent to the members of the several societies. Members of the Museum on making request of the Director will be provided with the Bulletin as issued.

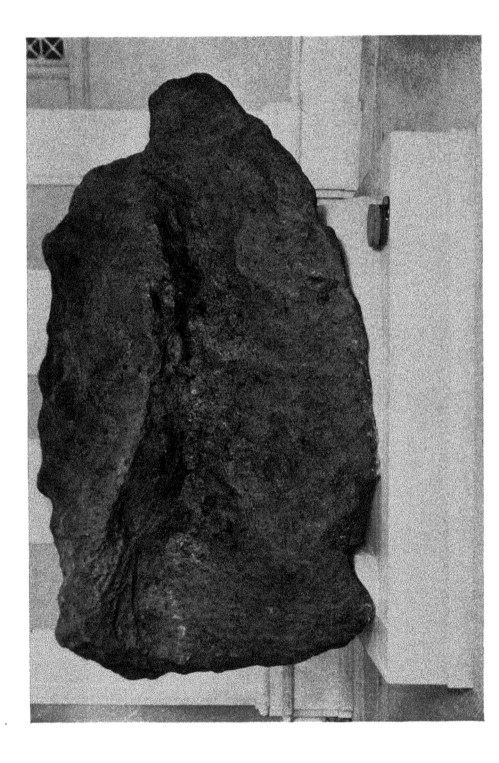

The American Museum Journal

VOL. VII   NOVEMBER, 1907   No. 7

AHNIGHITO, THE GREAT CAPE YORK METEORITE.

AST month mention was made in the JOURNAL of the removal of the Great Cape York (Greenland) Meteorite "Ahnighito" from the archway in front of the building to its permanent resting place in the Foyer. This month we present as our frontispiece an illustration showing the iron in position. Like the Willamette meteorite on the other side of the entrance, it rests on a solid pedestal of concrete which has been built up through the floor from the rock beneath the cellar. Thus the supports of these heavy masses of iron are entirely independent of the building, and no jarring can cause them to threaten the safety of the structure.

COLLECTIONS FROM THE CONGO.

HROUGH negotiations recently carried on with the Belgian authorities, arrangements have been made to establish in the Museum an extensive exhibition illustrative of the ethnic and zoölogical conditions of the Congo region in Africa. The proposition to establish such a permanent exhibition in New York City has appealed so strongly to King Leopold that he has authorized the government of the Congo Free State to open a special credit to aid in its formation. Two large shipments comprising more than 1500 specimens have already been received.

It is planned to devote two halls of the new wing of the American Museum, now in process of construction, to African ethnology. The Congo section will be beautifully decorated with frescoes showing the scenery and with real examples of the flora and fauna of the country, so that the specimens illustrating human life will appear in surroundings appropriate to the different tribes they represent. Some of the groups will be reproductions of the magnificent life-like figures in the Tervueren Museum, near Brussels, and all will have the appearance of animation, as they will portray the natives engaged in hunting and in their ordinary peaceful occupations. When the great number and diversity of the tribes in Central Africa, both in the valleys and on the uplands are considered, the scope for picturesque display may be realized.

In addition to the ethnological specimens, M. Liebrechts, secretary general of the Congo Department of the Interior, has promised the most complete data possible in the shape of photographs, statistical documents and samples of exported products, together with the entire series of the scientific publications of the Congo Independent State.

On its part, the American Museum of Natural History will send out expeditions to procure specimens illustrating fully the animal and plant life of this region. It will also collect all available data from independent trustworthy sources, concerning the Congo State, its discovery, history, resources and administration, so that all sides of the question will be presented for the examination of those interested.

The continent of Africa is being so rapidly opened up to occupation by civilized peoples that it is of the highest importance, from the point of view of the ethnologist, that collections be made without delay illustrating the life and the history of the savage and semi-savage tribes now living there. Hence the material recently received from the Congo, together with that which has been promised and that which had been obtained from East Africa and elsewhere will form an extensive and comprehensive collection which will probably, in a comparatively few years, be unique and of inestimable value.

A COLLECTION FROM THE TUKÁNO INDIANS OF SOUTH AMERICA.

HE Museum acquired in September a large amount of ethnological material from the Tukáno Indians of South America, the result of an eight-month sojourn of the well known scientist and collector, Mr. Hermann Schmidt, among the almost unknown people of the Rio Caiarý-Uaupes, a tributary of the Rio Negro in the State of Amazonas, Brazil. The locality is so remote from civilization, and the difficulty and danger incurred in reaching it are so great, on account of the numerous waterfalls and rapids of the rivers, which are the only highways, that many of the inhabitants had never seen a white man before Mr. Schmidt's arrival, and their mode of life and customs have probably changed but little since the beginning of the historic period in South America.

The primitive condition of the Tukáno Indians gives particular value to this collection of their household utensils, implements of war and the chase, clothing, ceremonial objects and ornaments, since these

SKETCH MAP OF NORTHWESTERN SOUTH AMERICA
The region occupied by the Tukáno Indians is along the Caiary-Uaupes River near the cross (+)

objects throw light on some pages, at least, of the history of the aborigines before the advent of white men.

By far the most striking specimens in the collection are the pieces of feather-work, of which there are about three hundred, consisting of a great variety of head-dresses, waist-bands, ornaments for the legs and arms and plumes to be carried in the hand. These ornaments are never worn except on ceremonial occasions, and then only by the men, the women wearing little clothing and but few ornaments. The feathers used in making these objects are largely from the red and blue macaw, various members of the parrot family, and a species of the heron. In looking at this collection of feather-work one is astonished at the delicacy and artistic beauty of the different objects and the surprising knowledge of color effects shown in the combinations of the feathers.

Some other notable objects in the collection are spears, shields, bows, arrows, blow-guns with their poisoned arrows, fish traps of basketry and numerous baskets in varied forms. Among musical instruments there are drums, rattles in many forms, pan-pipes and whistles made of deer and jaguar bones. A series of curious specimens illustrates the method of smoking the native tobacco. A cigar from ten to fifteen inches long and about an inch in diameter is made by rolling tobacco in a wrapper of bark and is fastened between the prongs of a wooden cigar-holder. The holder, which is about two feet long, exactly resembles a tuning fork in shape, except that the handle is longer and is sharpened to a point. After lighting the cigar, the Indian sticks the sharp end of the holder into the ground and lies at ease in his hammock, reaching out from time to time to draw in a whiff of smoke from the big cigar.   C. W. Mead.

THE PRONGHORN, OR AMERICAN ANTELOPE.

HE Pronghorn (*Antilocapra americana* Ord) belongs to a family of its own, combining features of the deer tribe with those of the goat and antelope, and is one of the most beautiful and interesting of our large game animals. It lives on the open rolling plains of the Western States, where at one time it was to be found in bands of hundreds or even thousands. No animal on the American continent compares with the Pronghorn in speed and keenness of vision, but, although very wary, the innate curiosity of the beast often leads to its destruction. A waving handkerchief or anything which excites its interest will frequently draw it within rifle-shot, and, like the buffalo, it has been reduced in numbers until only a few scattered herds remain.

A peculiar feature of the Pronghorn is its two white rump-patches which may be raised or lowered at will and are used as signals, for on a bright day the disc gives flashes of light which can be seen at a long distance. This odd habit, called "flashing," is illustrated in the photographs and by the young male in the group. Unlike other hoofed animals, the horns of the antelope are placed directly above the eye, and

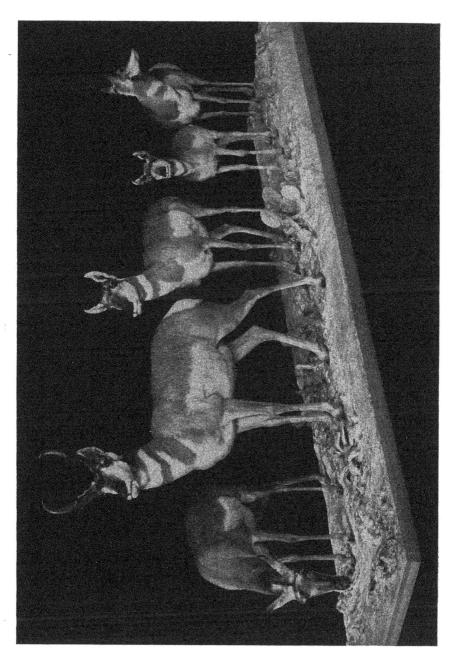

PRONGHORN, OR AMERICAN ANTELOPE
Group in East Mammal Hall, No. 207 of second floor

have a bony core from which the outer covering is shed annually.   The
core itself always remains in place.

THE NEW HALL OF RECENT FISHES.

 HALL devoted to the exhibition of Fishes has just
been opened to the public on the second floor of the
Museum, at the extreme end of the north wing.   The
collection consists of a fine series of mounted skins,
casts and skeletons, supplemented by colored plates
of various typical and striking forms.   It is intended
(1) to give a synoptic view of the Fishes of the world based on anatomical characteristics; (2) to emphasize, by means of descriptive labels,
the food-fishes and others having commercial value or striking peculiarities; (3) to bring into relief certain general features of biological
significance, like variation, apparent degeneration, protective mimicry,
brilliant coloration of tropical forms, and adaptation to such special
conditions as food-supply and deep-sea life.

A large study-collection of alcoholic specimens supplements the
exhibition series.   This material is in one of the Museum laboratories,
where it is available for research work by students.

THE FOREL COLLECTION OF ANTS.

PROFESSOR AUGUSTE FOREL of Yvorne, Switzerland, has just contributed an important addition to the large collection of Formicidæ
already in the Museum by donating a series of some 3,500 specimens,
representing about 1,400 species of exotic ants.   In this collection there
are nearly 800 type specimens, which are, of course, invaluable to future
students.   The Museum collection is now so extensive and contains so
many of the 5,000 known species, subspecies and varieties of these
highly variable insects that the study of additional materials from any
part of the world can be undertaken here profitably and without expend-
ing time in going over the much scattered and often very inadequate
descriptions of some of the earlier myrmecological writers.

..g pec...
reatures of biological
.eration, protective mimicry,
nd adaptation to such special
. . life.
coholic specimens supplements the
in one of the Museum laboratories,
k by students.

V OF ANTS.

....zerland, has just con-

PRECIOUS

George

The specimen, which is twice the size of the il
and one half inches wide, was found as a

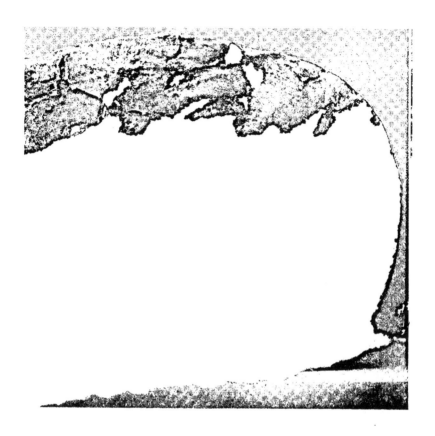

three quarters inches long by six
ew South Wales, Australia.

THE CIRCULATING NATURE STUDY COLLECTIONS.

URING the school year 1906–1907 the circulating Nature Study Collections were used by 660,740 pupils in the city public day schools and 79,358 pupils in the vacation schools. The demand for the sets has increased so that the Museum last summer procured an automobile to facilitate delivery to the schools. The specimens in the collections have been selected with a view to supplying material required in the Syllabus of Nature Study issued by the Board of Education, and there are now thirteen of these collections, containing in all five hundred sets. The collections and the number of pupils in the ordinary schools using them are shown in the following list:

CO LECTIONS.

NO. OF PUPILS.

Native Birds. Adapted for Grades 1A–4B:
Owl Set — Containing owl, chickadee, nuthatch, song
    sparrow, kinglet 89,565
Blue Jay Set — Containing blue jay, woodpecker, cross-
    bill, junco, English sparrow 85,882
Robin Set — Containing robin, red-winged blackbird,
    oriole, meadow-lark, chipping-sparrow 76,813
Bluebird Set — Containing bluebird, phœbe, barn swal-
    low, house wren, chimney swift 66,394
Tanager Set — Containing scarlet tanager, red-eyed vireo,
    goldfinch, humming-bird and pigeon 64,957
Insects. Adapted for Grades 2A–5A:
Containing cynthia and cecropia moths, monarch butter-
    fly, etc., and typical representatives of the different
    groups of insects 41,327
Special Insects. Adapted for Grades 2A–5A:
Containing life-history of cecropia moth, development of
    monarch butterfly, life and work of honey-bee and
    household insects 81,285
Mollusks. Adapted for Grades 4A and 5A:
Containing shells of about twenty-five mollusks, including
    the oyster, clam and chambered nautilus . . . 26,325
Crabs. Adapted for Grade 5A:
Containing relatives of the common blue crab . . 1,830

CO .LECTIONS.

Starfishes and Worms. Adapted for Grades 4A and 5A:
   Containing typical species of the two groups .   .   . 24,848
Sponges and Corals. Adapted for Grades 4A and 5A:
   Containing examples of about fifteen species .   .   . 14,501
Minerals and Rocks. Adapted for Grades 3B and 4A:
   Containing twenty specimens of minerals and building
   stones 48,816
Native Woods. Adapted for Grades 2A and 5B:
   Containing elm, hickory, maple, white birch, ailantus,
   sweet-gum, sour-gum, chestnut, sycamore. Speci-
   mens show cross, longitudinal and oblique sections
   of the wood, characteristic bark, annual ring, etc.   38,197
Total number of day-school pupils   660,740

The number of schools using the collections from September, 1906, to June, 1907, was 273, distributed among the boroughs as follows:

Manhattan,	125	Bronx,	24
Brooklyn,	73	Queens,	11
Richmond,	13	Corporate Schools,	27

The increase in the use of the collections by the pupils of the vacation schools has been marked and gratifying. Twenty-seven centers used the collections during July and August, 1907.

The work of the Museum in furnishing these collections to the public schools has attracted wide attention not only in this country but also in Europe, and several cities, particularly Newark, Milwaukee and St. Louis, have taken steps to introduce similar collections into their schools.

THE MODEL OF THE ATLANTIC SQUID.

HE Department of Invertebrate Zoölogy has recently placed on exhibition, in the Synoptic Hall, No. 107 of the ground floor, an enlarged model of the Common Squid *(Loligo pealii)*, a marine invertebrate common off the Atlantic Coast especially about Woods Hole, Massachusetts, where it is very destructive to the herring fisheries. The model, an illustration of which is given on page

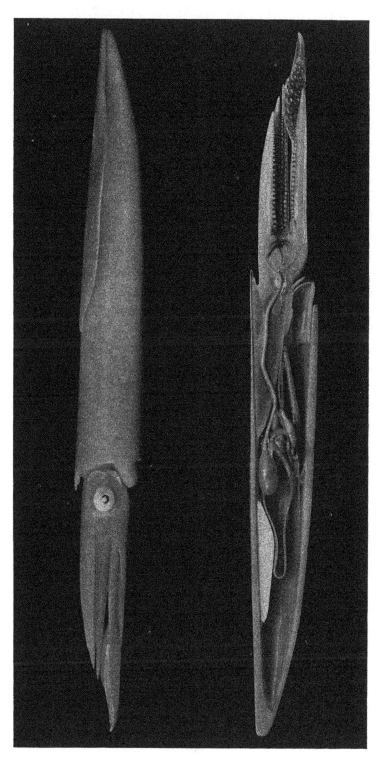

MODEL OF THE COMMON ATLANTIC SQUID

Two views showing external features and internal anatomy.   Synoptic Hall, No. 107 of the ground floor

The illustration is about one fifth as long as the model

115, was prepared in the Museum under the direction of Dr. L. W. Williams, and represents the male cut so as to show the peculiar internal anatomy of the animal. It is about twice (linear) natural size.

The Squid belongs to the Decapods, a subdivision of the Mollusks, and is closely related to the Octopus. The species represented by the model averages more than a foot in length and is somewhat cigar-shaped. It has a distinct head, furnished with a single pair of eyes which are the most perfect found among invertebrates and closely resemble the vertebrate eye itself, though originating in the embryo in a quite different way. The head is provided with a parrot-like beak, especially adapted for tearing flesh. The beak is set in a circular mouth surrounded by ten flexible arms, or tentacles, eight of which are of equal length and studded with suckers for grasping the animal's prey. The remaining two tentacles are much longer than the others and are without suckers, except on the club-shaped ends. The surface of the body is entirely covered with clusters of reddish pigmented spots known as chromatophores. These are ordinarily pale pink, but when the animal is excited, they become a deep red.

Most animals of this group are provided with an "ink-sac" which, in the case of the Cuttlefish (*Sepia officinalis*), secretes an intensely black fluid, the sepia of commerce. When the squid is startled, the "ink" is forcibly ejected from the body through a duct opening into the anus. The ink mixes with the surrounding water to form a black cloud, under cover of which the animal escapes.

The mode of progression of the Octopus and Squid is unique in the animal kingdom. The body proper is inclosed on the lower side in a "mantle cavity" to which water is freely admitted. At the entrance to this cavity is a flexible funnel or siphon with the small end pointing outward and forward. By filling the mantle-cavity with water and forcibly ejecting it through this funnel the animal is shot backward like a sky-rocket and at remarkable speed. When going forward, the mouth of the funnel is bent so as to shoot the stream of water backward.

MUSEUM NEWS NOTES.

THE permanent endowment fund of the Museum has recently been increased by a gift of $10,000 from Mrs. J. B. Trevor, and by the payment of a bequest of $25,000 from the estate of William P. Davis, Esq.

THE Department of Invertebrate Palæontology has recently placed on exhibition several interesting series of fossils. Among these may be mentioned about fifty specimens of Crinoids, or "sea lilies," from the famous beds of Lower Carboniferous age at Crawfordsville, Indiana. These are part of a particularly fine set that was received with the Cope Collection, presented to the Museum by President Jesup. Another noteworthy collection is the series of Unios from Hell Creek, Montana, presented to the Museum by Mr. Barnum Brown. These Unios are of latest Cretaceous time, that is to say they are millions of years old, but they are so much like the "fresh water clams" which inhabit the rivers and lakes of the Mississippi basin at the present time that the region in which they occur is undoubtedly the original home of our living forms. The present Unios are the shells from which are obtained the fresh water pearls of commerce. A little series of fossils from Grantland, 82° 37' north latitude, was brought back by the Peary Expedition in 1906. The fossils are of Carboniferous age and prove the existence of a mild climate in these far northern regions in Palæozoic time.

DR. D. LE SOUËF, Director of the Zoölogical Gardens, Melbourne, Australia, and delegate of the Colonial Government to the Seventh International Zoölogical Congress, gave an illustrated lecture at the Museum on Monday evening, September 9, under the auspices of the New York Academy of Sciences. His subject was "The Wild Animal Life of Australia," and he presented a series of remarkable and interesting photographs illustrating the strange animals of Australia and their home surroundings. Dr. Le Souëf's collection of such photographs was made by himself and is a result of wide experience in the field. It is considered the most complete in existence.

PROFESSOR MARSHALL H. SAVILLE, Honorary Curator of Archæology, arrived in New York September 10 on his return from Ecuador, where he had devoted about three months to field work in the coast region. He had most excellent success in collecting, and obtained a large quantity of material illustrating the Pre-Columbian life of the region, a region which is practically unknown to the scientific public. Mr. George H. Pepper of the Department of Ethnology, who accompanied Professor Saville, remained in Ecuador for additional work and returned to New York October 2.

PROFESSOR HENRY E. CRAMPTON of Columbia University made a second expedition to the Island of Tahiti during the past summer in behalf of the Museum, carrying on the studies which he began last year with regard to the effect of geographical isolation as a factor in specific evolution and the determination of data relating to the inheritance of sundry specific characters.

DURING a trip to Germany last summer Professor Bashford Dean, Curator of Fossil Fishes, secured for the Museum five beautiful specimens of fossil fishes from the celebrated lithographic stone quarries at Solnhofen, Bavaria. Among these are a *Propterus*, a *Caturus* and a *Megalurus* that are hardly excelled by any similar specimens in the German museums.

H. W. SETON-KARR, ESQ., of Wimbledon, England, has presented to the Department of Archæology a splendid series consisting of seventy-one specimens of palæolithic implements collected by him in the Districts of Poondi and Cazeepet, Madras Presidency, India. These implements are of red argillaceous sandstone and were washed out of Pleistocene alluvial deposits containing quartzite boulders. These relics of the early Stone Age have been placed on exhibition in the gallery cases of the Peruvian Hall, No. 302 of the gallery floor.

A LARGE ethnological collection made in Korea in 1906 and 1907 by Dr. C. C. Vinton has been presented to the Museum. The material consists principally of vessels of glazed and unglazed pottery, many of which are of beautiful design and finish.

THE collection illustrating the culture of the Indians of the Plains has been enriched by the accession of two decorated buffalo robes from the Sioux tribe. Since the practical extermination of the buffalo twenty-five years ago such robes have become extremely scarce.

THE Department of Archæology has received from Mr. Alanson Skinner a series of specimens collected for the Museum this year in Ontario, Livingston and Erie Counties, New York, from sites formerly occupied by the Seneca and Neutral Indians of Iriquoian stock.

LECTURE.ANNOUNCEMENTS.

MEMBERS' COURSE.

The first course of illustrated lectures for the season 1907–1908 to Members of the American Museum of Natural History and their friends will be given according to the following programme:
Thursdays at 8:15 P. M.

November 7.— FRANK M. CHAPMAN, "Bird Studies in the Bahamas, South Atlantic States and Northwestern Canada."
November 14.— HENRY FAIRFIELD OSBORN, "The American Museum Expedition to the Fayoum."
November 21.— EDMUND OTIS HOVEY, "A Month's Tour of the Yellowstone Park."
December 5.— HARLAN I. SMITH, "An Unknown Field in American Archæology."
December 12.— F. A. LUCAS, "The Fur Seal; its History and Habits."

PUPILS' COURSE.

Open to School Children, when accompanied by their Teachers, and to Children of Members, on presentation of Membership Ticket.
Lectures begin at 4:00 P. M.

	Oct.	Nov.
Monday,	28	18.— "Among the Filipinos." By G. H. Sherwood.
Wednesday,	30	20.— "The Panama Canal." By E. O. Hovey.
	Nov.	
Friday,	1	22.— "Our Native Birds and Their Habits." By F. M. Chapman.
Monday,	4	25.— "Early Days in New York City." By R. W. Miner.
	Dec.	
Wednesday,	6	4.— "Forests and their Dependent Industries." By A. C. Burrill.
Friday,	8	6.— "Historic Scenes in New England." By G. H. Sherwood.
Monday,	11	9.— "Peoples of the Earth." By H. I. Smith.
Wednesday,	13	11.— "Scenes in Our Western States." By R. C. Andrews.
Friday,	15	13.— "Famous Rivers of the World." By R. W. Miner.

LEGAL HOLIDAY COURSE.

Open free to the public. No tickets required.
Thanksgiving Day, November 28, 3:15 P. M. EDMUND OTIS HOVEY, "The Yellowstone National Park."

PEOPLE'S COURSE.

Given in coöperation with the City Department of Education.

Tuesday evenings at 8 o'clock. Doors open at 7:30.
A course of lectures illustrated with stereopticon views.

November   5.— Professor Char.es L. Bristol, "The Bermudas."
November 12.— Edwin E. Slosson, Ph. D., "The Panama Canal."
November 19.— Professor Henry H. Rusby, "The Delta of the Orinoco."
November 26.— Miss Carolina H. Huidobro, "Typical Life in Chili."

Saturday evenings at 8 o'clock. Doors open at 7:30.
A course of three lectures by Professor Samuel C. Schmucker.

November   2.— "Little Brothers of the Air (Birds).'
November   9.— "Modern Mound Builders (Ants)."
November 16.— "My Foster Children (Animals as Pets)."
November 23.— J. Russell Smith, Ph. D., "The Story of a Steel Rail."
          The first of a course of four lectures on "Commercial
          Geography" illustrated by stereopticon views.
November 30.— "The Story of a Ton of Coal."

Children are not admitted to these lectures, except on presentation of a
Museum Member's Card.

MEETINGS OF SOCIETIES.

Meetings of the New York Academy of Sciences and its Affiliated Societies
will be held at the Museum during the current month as follows:

On Mondays at 8:15 P. M.   The New York Academy of Sciences:
  November 4.— Business meeting and Section of Geology and
      Mineralogy.
  November 11.— Section of Biology.
  November 18.— Section of Astronomy, Physics and Chemistry.
  November 25.— Section of Anthropology and Psychology.

On Tuesday evenings as announced:
  The Linnæan Society, The New York Entomological Society and
  The Torrey Botanical Club.

On Friday evenings as announced:
  The New York Microscopical Society.

Full programmes of the meetings of the several organizations are published in the weekly *Bulletin* of the Academy and sent to the members of the
societies.   On making request of the Director of the Museum, our Members
will be provided with this *Bulletin* as issued.   The meetings are public.

The American Museum Journal

Vol. VII   DECEMBER, 1907   No. 8

A REPORT ON EXPEDITIONS MADE IN 1907 UNDER THE "NORTH AMERICAN ORNITHOLOGY FUND."

HROUGH the continued generous coöperation of the subscribers to the "North American Ornithology Fund," further important additions have been made to our series of "Habitat Groups" of North American birds.

The collections and field studies on which these groups are based can be made only during the nesting season. The work for this year was therefore planned to cover as long a nesting period as possible, beginning with southern species which nest as early as January and ending with northern birds which are not concerned with domestic affairs until July. In brief, the schedule was as follows:

March.— Southeastern Bahamas for Man-o'-war Birds and Boobies.

April.— Southern border of the Florida Everglades for Spoonbills and Ibises.

May.— South Carolina for White Egrets.

June.— Plains of Saskatchewan for Wild Geese and Grebes.

July.— Summits of the Canadian Rockies for Ptarmigan and other Arctic-Alpine Birds.

The species of birds here included show wide variation in form and in nesting habits, while the country in which they live — their "habitat" — presents an even greater diversity, as we pass from a coral islet to a mangrove swamp or a cypress forest and over the rolling plains to the snow-clad mountain crests. The subjects selected were thus designed to add to the zoölogical as well as geographical instructiveness of the exhibits as a whole.

A series of mishaps so prolonged the Bahaman expedition that I was prevented from reaching the Everglades in time to find Spoonbills nesting, but, with this exception, the schedule outlined above was followed with eminently satisfactory results.

On March 28, with Dr. Alfred G. Mayer and Mr. George Shiras, 3d, I sailed from Miami, Florida, for Nassau, Bahamas, aboard the 58-foot auxiliary ketch, "Physalia," belonging to the Marine Biological

Laboratory of the Carnegie Institution. Dr. Mayer, who is the Director of the laboratory, was in command, and to his coöperation the Museum is indebted for the success which attended our efforts to secure material and studies for the groups of Man-o'-war Birds and Boobies; indeed, had it not been for Dr. Mayer's skilful seamanship, it is probable that the expedition would not have returned at all.

Nassau was reached March 29 at midnight. Laboratory supplies were landed for the use of members of the staff who proposed to remain

CAMP ON CAY VERDE

here to study, and, permission to collect the birds needed having been promptly granted by the Bahaman Government, we set sail for Cay Verde, March 31 at 7 A. M.

Cay Verde is an uninhabited islet some forty acres in area situated on the eastern edge of the Columbus Bank, between the Ragged Islands and Inagua. It is only 250 miles from Nassau, but adverse weather conditions, which at times threatened us with serious disaster, lengthened our voyage thither to ten days. The absence of definite information, both as to the number of birds frequenting Cay Verde and the time of

THE WILD-TURKEY HABITAT GROUP

In process of construction

their nesting, made the outcome of our trip more or less uncertain, and the difficulties encountered in reaching this remote islet added in no small degree to the pleasure with which we found it thickly populated with Boobies and Man-o'-war Birds whose nesting season was at its height. There is no harbor at Cay Verde, and, fearing that we might be forced by a storm to sail before our work was finished, Mr. Shiras and I camped on the islet, while Dr. Mayer anchored off shore, changing his position from one side of the Cay to the other as the wind shifted.

We estimated that there were about 3000 Boobies and 500 Man-o'-war Birds on Cay Verde. The Boobies nested on the ground, the Man-o'-war Birds in the dense thickets of sea grape and cactus. Some nests contained fresh eggs, but the larger number held young birds in various stages of development, while a few young were already on the wing. The existing conditions therefore presented an epitome of the whole nesting season.

The Boobies were remarkably tame, and our intrusion occasioned surprise and resentment rather than fear. One could walk among them as one would through a poultry yard, examining the nests and their occupants without attempt at concealment. The Man-o'-war Birds were more suspicious but still were approached without difficulty. Under these circumstances photographs and specimens were easily secured, and in the course of three days satisfactory material was collected for the proposed group. A much longer period would be required to make adequate studies of the life of this bird community. Cay Verde was left April 11, and, after encountering the usual unfavorable conditions and some mishaps, we arrived at Miami, Florida, April 29.

It being now too late to do the work planned for southern Florida, I proceeded to South Carolina, being joined by Mr. J. D. Figgins of the Museum's Department of Preparation and by Mr. Bruce Horsfall, the artist who has so successfully painted many of the backgrounds of the groups already completed.

It has long been our desire to include the White Egret in the series of "Habitat Groups," but plume hunters have brought this bird so near the verge of extermination that our efforts to find a "rookery" in which suitable studies might be made were fruitless before the present year. In February, 1907, information was received of the existence of a colony of Egrets on a large game preserve in South Carolina, and the owners of the preserve readily granted the Museum permission to make the necessary studies and collections. On the arrival of our expedition every facility for our work in the way of transportation, guides and other necessaries was accorded us.

When the ground in which the rookery is situated was acquired by the

CYPRESSES IN WHICH THE EGRETS NEST

The blind from which the birds were studied may be seen in the upper right-hand corner of the picture

club now owning it, plume hunters had nearly exterminated the aigrette-bearing Herons which formerly inhabited it in large numbers. A few had escaped, and after seven years of protection they have formed one

THE BROWN PELICAN HABITAT GROUP

of the largest colonies of this much-persecuted bird now existing in the United States. Six other species of Herons were found nesting with the White Egrets, the whole forming a rookery like those which existed commonly in the days of Audubon, but which are now almost unknown in the United States.

A former "plumer," now chief warden in charge of the preserve, stated that both the Little White or Snowy Egret and the Roseate Spoonbill were once found in the region, but their complete annihilation left no stock which, under protection, might prove the source of an ever-

AN EGRET FAMILY

increasing progeny. It is doubtful if these birds could be introduced, but in any event the preservation of the White Egret alone is a sufficient cause for thanksgiving, and bird lovers will learn with gratification of the existence of an asylum where this beautiful creature will long be assured of a haven of refuge.

The Egrets were nesting high in the cypress trees which grow in a lake several miles in length. In order, therefore, to make the photographic studies so essential to the taxidermist in securing life-like poses. for his subjects, as well as to learn something also, of the Egrets' little-known home life, the artificial umbrella-blind employed on many previous occasions for similar purposes was placed fifty feet up in a cypress tree and draped with "Spanish moss" (*Tillandsia*). From it photographs of the birds nesting in neighboring trees were eventually made. The surroundings were of great beauty, and Mr. Horsfall's carefully made studies will enable him to reproduce in his background the singular charm of a flooded cypress forest.

RING-BILLED AND CALIFORNIA GULLS

On June 5, accompanied by Mr. L. A. Fuertes, as artist, I left New York for Maple Creek, Saskatchewan, on the line of the Canadian Pacific railway. This is a region of rolling plains dotted with lakes and ponds which, when the water is not too alkaline, support, in their shallower parts, a dense growth of rushes, the home of Grebes, Coots, Bitterns, Franklin's Gulls and Ruddy, Red-headed and Canvasback Ducks. About the grassy borders of the lakes and sloughs, Mallards, Gadwalls, Pintails, Widgeon, Blue-winged Teal and other ducks nest. These

HABITAT GROUP SHOWING CACTUS-DESERT BIRD LIFE OF ARIZONA

Hall No. 308, Gallery Floor

species were also found on islands in the lakes, where alone the Wild Goose was known to nest, while some small islets were virtually covered by hosts of Gulls and Pelicans.

On the prairies Long-billed Curlew, Marbled Godwits, and Bartramian Sandpipers or "Upland Plover" as sportsmen know them, lay their eggs. The region has well been called the nursery of wild-fowl, as at one time were our border states to the south; but the advance of civilization, which first transforms a buffalo range into a cattle country and

CAMP AT PTARMIGAN PASS

later into a wheat ranch, has already reached the early stages of its agricultural development about Maple Creek and the forced retreat of the wild fowl to the more remote north is only a question of time. The Canadian Government would do well to set aside some of its still unsettled lands as permanent breeding reservations to which year after year the water-fowl could return to nest. Such reservations would be nurseries and, by permitting a bird to reproduce, would be of infinitely more importance than preserves which afford protection only during the winter.

Near Maple Creek materials were secured for groups of Wild Geese, Western and Eared Grebes, the Long-billed Curlew and Bartramian Sandpiper, due permission having first been received from the Chief Game Guardian of the Province. The lack of timber and of drinking water made this region poor camping ground and while hunting and collecting we were given quarters with Mr. Andrew Scott on Crane Lake and with the Messrs. Baynton on Big Stick Lake. To these gentlemen

A PTARMIGAN ON HER NEST

Mr. Fuertes about to stroke the bird

we are indebted not alone for entertainment but also for much practical assistance.

July 2 we resumed our western journey, in search now of those Arctic birds which on the alpine summits of the Rocky Mountains find congenial surroundings. After inquiry at various places, we decided to camp near the Ptarmigan Lakes, where we were informed the birds we wanted could be found. Saddle and pack horses and a guide were secured at Laggan, and on July 8 we encamped just below the entrance to Ptarmigan Pass near timberline, which here is at an altitude of 7500 feet above the sea.

The alpine spring was at its height. The wet meadows, from which the snow had but lately disappeared, were yellow with butter-cups, the borders of the rapidly shrinking snow banks were starred with large white alpine anemones, on the drier slopes heather bloomed luxuriantly, and the rocks were covered with flowering Dryas. The lakes were still ice-bound; the mercury reached the freezing point nightly, and we experienced several storms of snow and sleet.

Our work in this indescribably picturesque region was unexpectedly successful, specimens of birds and plants and a large number of photographs being obtained. Furthermore, the view from the heather-grown home of the Ptarmigan, which will form the actual foreground of our group, southward through the Ptarmigan Pass was of exceptional grandeur, even in this land of sublime scenery. The successively fainter timber-clad shoulders of the gap leading to the Bow Valley are backed by Mt. Temple towering impressively, the central peak on horizon marked to the east by the spire-like summits of the mountains about Moraine Lake and to the west by Mts. Hungabee, Lefroy and Victoria.

The tourists who climb these mountains or penetrate the valleys lying between them may obtain a far more striking view of the range by crossing the Bow River at Laggan and ascending the mountains to the north in which the studies for our Ptarmigan group were made.

FRANK M. CHAPMAN.

THERE has recently been installed in the Hall of Invertebrates, on the ground floor of the Museum, an interesting series of models showing the larval, pupal and adult stages in the life history of the mosquito which has been proven to cause the spread of malaria. The models are 75 times as great in linear dimensions as is the insect itself, and therefore the volume is more than 420,000 times that of the living animal. On this scale, the adult mosquito stands one and one half feet in height and is three feet long. The spread of the wings is three feet and the mouth parts (beak) are one foot long. All the details of the anatomical structure of the animal have been reproduced with scientific accuracy and painstaking labor. A guide leaflet upon this model and the life history of the malaria mosquito is in course of preparation.

AN ALEUTIAN BASKET.

HROUGH Mrs. Mary Graham Young the American Museum has recently received from the Women's National Relief Association, or Blue Anchor Society, a small Aleutian basket which is unique in its way. The Aleuts, the Indians dwelling on the Aleutian Islands, are considered a division of the Eskimo. Though now far advanced toward civilization, they still retain their native skill in the textile art. Their basketry first came to notice through specimens brought from the inhabitants of Attu Island, hence all baskets of this type usually pass under the name Attu. Taking the recent gift as a representative basket, it may be interesting to know something of its making.

The gathering and preparing of the material used in the weaving is an annual task of no small importance. Early in July a great harvesting party of women starts out to get the wild rye, called "beach grass," which is the only suitable basket material to be found on the islands. This is a coarse grass, with leaves about two feet long and half an inch wide, growing plentifully along the coast and on the hills. It is to the high land, however, that the women go to find the best material. Only two or three of the young and delicate leaves are selected from each stalk, hence this is no easy and rapid gathering, and it is with the greatest patience that the native women reap their harvest.

The drying or curing process is a long one. First the beach grass is spread out in rows on the ground in a shady place. As it dries, it is turned frequently. This stage of the drying takes about two weeks. Then the grass is sorted as to size and taken into the house, the coarser leaves split with the thumb-nail into three parts, the middle or midrib being discarded, and the fine leaves left unsplit, because still too tender for such treatment. Bundles are now made of all the material, and for a month on cloudy days these are hung out on a line. The final drying is done indoors and then the grass is separated into small wisps about the size of a finger, with the ends braided loosely so that the grass may not tangle when a thread is pulled from the wisp. This single thread may be split by the thumb nail to any size desired at the time it is used.

A great part of the weaving is done during the winter months and indoors. The underground huts of the Aleuts are made of driftwood, wreckage, or timber deposited by ships, covered over with sod. These grass-covered mounds, which are about six feet high, have a little door at one end and a small glazed window at the other, and it is marvelous

A.EUTIAN BASKET

that such fine and beautiful baskets can be made in such places, and with such light, the specimen here illustrated having from twenty-five to thirty stitches to the inch. John Smith's Indians used to suspend their baskets from the limb of a tree during the weaving, but these people hang theirs from a pole after the bottom has been completed or support them on sticks thrust into the ground, weaving the sides downward, that is with the basket upside down.

The Aleuts make several weaves. This basket is of plain twine-weave with two exquisitely wrought rows of hemstitching on the bottom, while the sides are decorated with four borders of false embroidery.

The encircling bands, composed of lines and rectangles are in red,
green and blue worsted and silk thread and white skin from the throat
of a fish of the sculpin family cut into fine threads.

On the lid are three more bands, with an attractive medallion in

A.EUTIAN BASKET. THE BOTTOM

colors on the knob or handle. The technique of the false embroidery
is interesting, as the patterns are woven into the texture, but not through
to the inside of the basket. At each stitch where the design is desired
as the two weavers which form the weft inclose the warp spoke, the outer
weaver is wrapped by the colored thread. Within the knob on the
cover are several pebbles that rattle when shaken. The sound is considered to resemble the rattling of stones on a beach as they are moved
by the waves. A full discussion of this subject can be found in the
Craftsman, March, 1904, and Mason's "Indian Basketry."

This exquisite basket, which is a masterpiece of Aleutian workmanship, was given to the Blue Anchor Society by the chief of the Attu in token of appreciation of certain assistance rendered the tribe during a period of distress, and it has been given to the American Museum for permanent preservation.   MARY LOIS KISSELL.

AN ANT-HUNTING TRIP TO EUROPE IN THE SUMMER OF 1907.

HE past summer was devoted by Dr. W. M. Wheeler to a study of the European Formicidæ (ants) and to securing extensive series of these insects for the Museum. This study was necessary for the purpose of throwing light on the structure, habits and relationships of our North American forms, many of which are regarded as mere subspecies or varieties of well-known palearctic (European and North Asiatic) species. After collecting for a short time at Ponta Delgada in the Azores, a locality in which no ants had been collected previously, at Gibraltar and at Genoa, Dr. Wheeler selected Switzerland as the best place in which to continue his work, because this country presents an extraordinary variety of physical and biological conditions and therefore an epitome, so to speak, of the whole ant fauna of Europe, and because it has become the classical locality for such studies through the remarkable work of Pierre Huber during the early years of the nineteenth century and of Auguste Forel during the past half century. Professor Forel, the most eminent of living myrmecologists, who until recently has been residing at Chigny near Lausanne on the shore of Lake Leman, showed the greatest interest in Dr. Wheeler's studies and gave him invaluable assistance, both by directing him to the most profitable localities in which to observe and collect ants and by accompanying him on excursions in the vicinity of Geneva and in Canton Vaud. Subsequently Dr. Wheeler made expeditions to Yvorne, Fully, Sion and Sierre in the upper valley of the Rhone (Canton Valais), to the Jura, to Locarno, Lugano, Bellinzona, Mendrisio, Monte Ceneri and Monte Generoso in Canton Ticino and to the Albula Pass, Samaden, Pontresina, St. Moritz and Silva Plana in the Upper Engadine (Canton of Grisons). Owing to the great variety and difference in the altitude of the country covered in these expeditions, it was possible to secure large series of specimens of all but a very few of the Formicidæ known to occur in central and northern Europe and to gain an intimate acquaintance with these insects and their parasites and messmates. Particularly valuable was the series of observations on the habits and development of the singular ant-nest beetle *Lomechusa strumosa* observed in the pine-woods of the Upper Engadine. After spending two months in Switzerland, Dr. Wheeler continued his observations at Würzburg in Bavaria and in the vicinity of Dresden in Saxony. In the former locality he was generously assisted in the collection of material by Professors Spemann and Lehmann of the University and in the latter by Professor Escherich of the Royal Academy of Forestry at Tharandt and Mr. H. Viehmeyer of Dresden. The results of Dr. Wheeler's comparative studies of the North American and European ant faunas will be published in the Museum " Bulletin."

TUESDAY afternoon, October 29, the National Association of Audubon Societies held its annual meeting at the Museum. The meeting was largely attended, and, after routine business had been despatched, it was addressed by Mr. Frank M. Chapman, Associate Curator of Birds, who gave an illustrated account of the "Home Life of the White Egret." Mr. William Dutcher was re-elected president of the organization.

THE first Conference of Anglers was held at the Museum Monday evening, November 11, under the presidency of Dr. Henry Van Dyke of Princeton University, for the purpose of exchanging views, obtaining information and uniting more closely and effectively for the protection of the game fishes and for the improvement of the sport of angling.

WEDNESDAY evening, October 30, the Museum, in coöperation with the New York Academy of Sciences, had the pleasure of offering to its members and their friends an illustrated lecture by Professor William Bateson of St. John's College, Cambridge, England. The lecture was upon the subject "The Inheritance of Color in Animals and Plants" and was a popular exposition of the now famous Mendelian Law of Heredity.

THE "FLYING-DRAGON" GROUP.

HE remarkable little lizard shown in the group illustrated on this page, the so-called Flying Dragon (*Draco volans* Gray), takes its name from the numerous wing-like membranous expansions of its sides, which act like aëroplanes as the animal jumps or floats from branch to branch of the trees in which it dwells. These

THE FLYING-DRAGON GROUP
In East Mammal Hall, No. 207 of the main floor

folds of skin are supported by the five or six posterior pairs of ribs and may be folded like fans. On the throat of the male are three pointed orange-colored appendages, of which the middle is the longer. In the female these appendages are blue. The metallic sheen of the body and the prettily marked orange and black wings of the animal harmonize perfectly with the surrounding foliage and the gaudily colored flowers among which it rests, and aid in concealing it from its enemies as well as from the insects for which it lies in wait. About twenty species of these Flying Dragons are widely distributed throughout the Indo-Malayan countries, though on account of their retiring habits they are nowhere considered common.

The six specimens in this group were collected in the Island of Nias, off Sumatra. The tree on which they perch is the Nutmeg (*Myristica fragrans*); the orchid is the beautiful *Phalænopsis schilleriana* of Indo-Malaysia, and the climbing vine is *Cissus discolor*, a member of the Grape family. The Butterfly is the graceful *Leptocircus curius*; the beetles are *Coryphocera dohrni* and an unidentified Buprestid of this region. The group was mounted by Mr. J. D. Figgins of the Museum staff.

MUSEUM NEWS NOTES.

AT a meeting of the Board of Trustees which was held at the Museum, November 11, resolutions of thanks were passed to the following friends of the institution:

To His Excellency C. A. M. LIEBRECHTS of Brussels, Belgium, for his assistance in connection with the Congo Exhibit, and he was elected a Patron.

To Professor A. FOREL of Yvorne, Switzerland, for his presentation of a collection of ants, and he was elected a Patron.

To Mrs. ROBERT WINTHROP for her contribution toward the development of the Habitat Groups of North American birds, and she was elected a Fellow.

To Mr. J. F. FREIRE MURTA for the gift of a collection of tourmalines, aquamarines and other gem material, and he was elected a Life Member.

To Mr. E. P. MATHEWSON for his gift of ethnological specimens from Chile, and he was elected a Life Member.

To Mr. FRANK K. STURGIS for his contributions to the field work of the Department of Vertebrate Palaeontology, and he was elected a Life Member.

To Mr. PERCY R. PYNE and Mr. J. P. MORGAN, Jr., for their contributions to the Alaskan Mammoth expedition.

To Mr. J. PIERPONT MORGAN for his gift of a boulder of New Zealand Jade.

At the same meeting His Honor, Mayor George B. McClellan and Hon. Herman A. Metz were elected Patrons, and the election of Messrs. F. L. St. John, William D. Guthrie and John Treadwell Nichols to Life Membership through the subscription of one hundred dollars each was announced.

Director Bumpus returned November 9 from a brief trip to Europe which was taken for the purpose of representing the American Museum at the dedication of the Senckenberg Museum at Frankfurt, Germany, October 13. While abroad, Dr. Bumpus took advantage of the opportunity to visit many of the principal museums on the Continent inspecting collections and museum methods in general.

Two pairs of antique carved elephant tusks and two carved ivory gods, all from the Benin district of the West Coast of Africa, have been acquired recently by the Museum. They were taken from King Prempeh by a detachment of the British army which had been sent into the country to punish his tribe for cannibalism. All the objects were veritable idols to which human sacrifices were made and which had been held in high veneration by the natives for generations.

The Museum has recently received forty-four interesting arrow heads found by H. W. Seton-Karr, Esq., during nine expeditions through the Desert of Fayoum, Egypt. Besides these, there are nine large arrow or spear heads, thirteen knives, three bent or wavy flakes worked on the edges and peculiar to the Fayoum Desert, one long worked flake and two adzes made of stone. The objects were found on the sites of ancient villages, but there is now no water near them, and the village-sites are indicated by mealing stones or grinders found bottom side up. There are also four celts or chisels from India.

Dr. Robert H. Lowie recently returned from a Museum expedition to Alberta and Montana. Dr. Lowie left New York on June 8 for Gleichen, Alberta, where he collected notes on the Northern Blackfoot. At Morley, Alberta, he camped for seven weeks with the "Stoney" Assiniboine, gathering a collection to represent their mythology. Then Crow Agency, Montana, was visited, where specimens and notes on the social and ceremonial organization of the Crow were obtained.

The bequest mentioned on page 116 should have been credited to the estate of Benjamin P. Davis, Esq.

LECTURE ANNOUNCEMENTS.

MEMBERS' COURSE.

Thursday evenings at 8:15 o'clock. Open to Members and to those holding complimentary tickets given them by Members.

December 5.— HARLAN I. SMITH, "An Unknown Field in American Archæology."

December 12.— FREDERIC A. LUCAS, "The Fur Seal — Its History and Habits."

PUPILS' COURSE.

Mondays, Wednesdays and Fridays, at 4 o'clock.

Open to School Children, when accompanied by their Teachers, and to children of members, on presentation of Membership Ticket.

Wednesday, December 4.— "Forests and their Dependent Industries." By A. C. BURRILL.

Friday, December 6.— "Historic Scenes in New England." By G. H. SHERWOOD.

Monday, December 9.— "Peoples of the Earth." By H. I. SMITH.

Wednesday, December 11.— "Scenes in our Western States." By R. C. ANDREWS.

Friday, December 13.— "Famous Rivers of the World." By R. W. MINER.

LEGAL HOLIDAY COURSE.

Fully illustrated. Open free to the public. No tickets required. Doors open at 2:45, lectures begin at 3:15 o'clock.

The programme for the season 1907–1908 is as follows:

Thanksgiving Day, November 28, 1907.

A Month's Tour of the Yellowstone Park . . EDMUND OTIS HOVEY

Christmas Day, December 25, 1907.

Hiawatha's People . . . . HARLAN I. SMITH

New Year's Day, January 1, 1908.

An Ornithologist's Travels in the West . . FRANK M. CHAPMAN

Washington's Birthday, February 22, 1908.

Mines, Quarries and "Steel Construction " . . LOUIS P. GRATACAP

PEOPLE'S COURSE.

Given in coöperation with the City Department of Education.
Tuesday evenings at 8 o'clock. Doors open at 7:30.
December 3.— Miss Caro ina H. Huidobro, " Typical Life in Chile."
December 10.— Mrs. M. Claire Finney, "The Land of the Incas."

Saturday evenings, at 8 o'clock. Doors open at 7:30.
December 7.— J. Russell Smith, Ph. D., "The Story of a Ton of Coal."
December 14.— J. Russe.. Smith, Ph. D., "The Story of a Piece of
Board."

MEETINGS OF SOCIETIES.

Meetings of the New York Academy of Sciences and its Affiliated
Societies will be held at the Museum during the current month as follows:

On Mondays at 8:15 P. M. The New York Academy of Sciences:

December 2.— Business meeting and Section of Geology and Mineralogy.
December 9.— Section of Biology.

On Monday, December 30.— The Linnæan Society.

On Tuesday evenings as announced:

The Linnæan Society, The New York Entomological Society and
The Torrey Botanical Club.

On Friday evenings as announced:

The New York Microscopical Society.

Full programmes of the meetings of the several organizations are published in the weekly *Bulletin* of the Academy and sent to the members of the
societies. On making request of the Director of the Museum, our Members
will be provided with this *Bulletin* as issued. The meetings are public.

Guide Leaflets Published by the
AMERICAN MUSEUM OF NATURAL HISTORY
For Sale at the Museum.

(*Issued as supplements to The American Museum Journal*)

No. 1.— THE BIRD ROCK GROUP. By FRANK M. CHAPMAN, Associate Curator of Mammalogy and Ornithology. October, 1901. *Price, 10 cents.*

No. 2.— THE SAGINAW VALLEY COLLECTION. By HARLAN I. SMITH, Assistant Curator of Archæology. December, 1901. *Price, 10 cents.*

No. 3.— THE HALL OF FOSSIL VERTEBRATES. By W. D. MATTHEW, Ph.D., Assistant Curator of Vertebrate Palæontology. January, 1902. *Out of print.*

No. 4.— THE COLLECTION OF MINERALS. By LOUIS P. GRATACAP, A. M., Curator of Mineralogy. February, 1902. *Revised edition, May, 1904. Price, 10 cents.*

No. 5.— NORTH AMERICAN RUMINANTS. By J. A. ALLEN, Ph.D. Curator of Mammalogy and Ornithology. March, 1902. *Revised edition, February, 1904. Price, 10 cents.*

No. 6.— THE ANCIENT BASKET MAKERS OF SOUTHEASTERN UTAH. By GEORGE H. PEPPER, Assistant in Anthropology. April, 1902. *Price, 10 cents.*

No. 7.— THE BUTTERFLIES OF THE VICINITY OF NEW YORK CITY. By WILLIAM BEUTENMULLER, Curator of Entomology. May, 1902. *Price, 15 cents.*

No. 8.— THE SEQUOIA. A Historical Review of Biological Science. By GEORGE H. SHERWOOD, A. M., Assistant Curator. November, 1902. *Price, 10 cents.*

No. 9.— THE EVOLUTION OF THE HORSE. By W. D. MATTHEW, Ph.D., Associate Curator of Vertebrate Palæontology. January, 1903. *Second edition, May, 1905. Price, 10 cents.*

No. 10.— THE HAWK–MOTHS OF THE VICINITY OF NEW YORK CITY. By WILLIAM BEUTENMULLER, Curator of Entomology. February, 1903. *Price, 10 cents.*

No. 11.— THE MUSICAL INSTRUMENTS OF THE INCAS. By CHARLES W. MEAD, Assistant in Archæology. July, 1903. *Price, 10 cents.*

No. 12.— THE COLLECTION OF FOSSIL VERTEBRATES. By W. D. MATTHEW, Ph.D., Associate Curator of Vertebrate Palæontology. October, 1903. *Price, 10 cents.*

No. 13.— A GENERAL GUIDE TO THE AMERICAN MUSEUM OF NATURAL HISTORY. Jan. 1904. *Out of Print.*

No. 14.— BIRD'S NESTS AND EGGS. By Frank M. Chapman, Associate Curator of Mammalogy and Ornithology. April, 1904. *Reprinted*, February, 1905. *Price, 10 cents.*

No. 15.— PRIMITIVE ART. July, 1904. *Price, 15 cents.*

No. 16.— THE INSECT-GALLS OF THE VICINITY OF NEW YORK CITY. By William Beutenmuller, Curator of Entomology. October, 1904. *Price, 15 cents.*

(Reprinted from The American Museum Journal.)

No. 17.— THE FOSSIL CARNIVORES, MARSUPIALS, AND SMALL MAMMALS IN THE AMERICAN MUSEUM OF NATURAL HISTORY. By W. D. Matthew, Ph. D., Associate Curator of Vertebrate Palæontology. Jan. 1905. *Price, 15 cents.*

No. 18.— THE MOUNTED SKELETON OF BRONTOSAURUS. By W. D. Matthew, Ph.D., Associate Curator of Vertebrate Palæontology. April, 1905. *Out of print.*

No. 19.— THE REPTILES OF THE VICINITY OF NEW YORK CITY. By Raymond L. Ditmars, Curator of Reptiles, New York Zoölogical Park. July, 1905. *Price, 15 cents.*

No. 20.— THE BATRACHIANS OF THE VICINITY OF NEW YORK CITY. By Raymond L. Ditmars, Curator of Reptiles, New York Zoölogical Park. October, 1905. *Price, 15 cents.*

No. 21.— THE DEVELOPMENT OF A MOLLUSK. By B. E. Dahlgren, D.M.D. January, 1906. *Price, 10 cents.*

No. 22.— THE BIRDS OF THE VICINITY OF NEW YORK CITY. By Frank M. Chapman, Associate Curator of Mammalogy and Ornithology. April–July, 1906. *Price, 15 cents.*

No. 23.— THE SPONGE ALCOVE. By Roy W. Miner, Assistant Curator of Invertebrate Zoölogy. Oct. 1906. *Price, 10 cents.*

(Published as a separate series.)

No. 24.— PERUVIAN MUMMIES. By Charles W. Mead, Department of Ethnology. March, 1907. *Price, 10 cents.*

No. 25.— PIONEERS OF AMERICAN SCIENCE. Memorials of the naturalists whose busts are in the Foyer of the Museum. April, 1907. *Price, 15 cents.*

No. 26.— THE FOYER COLLECTION OF METEORITES. By Edmund Otis Hovey, Associate Curator of Geology. December, 1907. *Price, 10 cents. In press.*

CHULPAS, OR BURIAL TOWERS
Sillustani, Peru

Peruvian Mummies

AND WHAT THEY TEACH

A GUIDE TO EXHIBITS IN THE PERUVIAN HALL

By CHARLES W. MEAD

DEPARTMENT OF ETHNOLOGY

NO. 24

OF THE

GUIDE LEAFLET SERIES

OF THE

AMERICAN MUSEUM OF NATURAL HISTORY

EDMUND OTIS HOVEY, EDITOR

New York. Published by the Museum. March, 1907

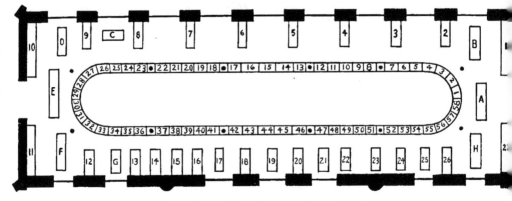

PERUVIAN HALL NO. 302.

Gallery Floor, West Wing.

PRESENT LOCATION OF THE OBJECTS DESCRIBED IN THIS LEAFLET.

	CASE
Mummy bundles	U 27
Mummies	U 27
Prayer sticks	R 4–5
Mummified animals	U 27
Trephined skulls	U 26
Skull Collection	U 26
Implements of war and the chase	U 21
Gold and silver objects	A
Baskets, mats and nets	R 17–18
Cloths	U 1
Materials and implements used in weaving	B
Quipus, or Record Fringes	R 1, 2
Coca leaves and outfit for chewing	R 11
Pottery	U 9, 10, 11, 12 and D, E, F
Chicha jars	On top of U 1, 3, 4, 5, 6, 7, 27
Collection from the West Indies	U 2
Musical Instruments	U 25

"U" refers to the upright cases; "R," to the railing cases.

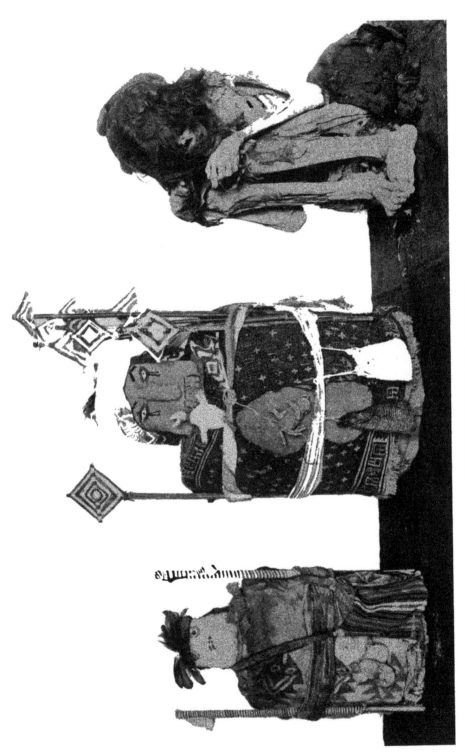

PERUVIAN MUMMY BUNDLES AND MUMMY

PERUVIAN MUMMIES AND WHAT THEY TEACH.

By Charles W. Mead,

Department of Ethnology.

NCIENT Peru, the land of the Incas, comprised not only the region included within the present Republic of Peru, but also the greater part of Ecuador, Bolivia and Chile and was about equal to that portion of the United States lying east of the Rocky Mountains. The Incas proper were a powerful tribe of warlike people inhabiting the great central plateau, from which dominating position they extended their conquest in all directions. They developed a much higher order of civilization than was found in other parts of the continent by the early European explorers, and the empire under their sway included many tribes speaking different dialects.

The history of the Ancient Peruvians must to a large degree be read in their graves, since they left no written records and the Spanish conqueror destroyed many of their cities and suppressed their customs. Like many other peoples the Peruvians bestowed much tender care on their dead, carefully preparing them for burial and placing with them in the ground many objects which were dear to them in life. Methods of burial are so intimately connected with the religious and other customs of a people that in the absence of other sources of information a study of graves or tombs may be expected to lead to important results. Fortunately for the archæologist, climatic conditions in some parts of Peru are such that "burials" have been well preserved. The region west of the Peruvian Cordillera, a narrow strip along the coast, is in the main a desert, the only fertile spots being the narrow valleys of the small rivers flowing down to the Pacific.[1] The tombs and graves are usually found on elevated places outside of the valleys where the extreme dryness of the air combines with the nitrous character of the sand, into which moisture has seldom found its way, to desiccate and preserve the bodies of the dead thus mummifying them naturally. The same factors have

Importance of the Burials

[1] The visitor is referred to the relief map of South America on the left as he enters the hall for a clear exposition of the topographic features of the region.

caused the clothing and objects placed with the dead to be preserved for many centuries.

As a rule the bodies were prepared for burial by placing them in a sitting position with the knees drawn up and the head and hands resting upon them, as is shown in the right-hand figure on page 6.   Sometimes,

Preparation for Burial   however, as appears from burials in the Chira Valley, in the extreme northwest of Peru, the body was extended at full length.   A few of the extended bodies have been found in other parts of the country, and two examples of this form from Surco, Peru, are in the collection.   After the body was placed in position, it was enveloped in wrappings of various kinds.   Sometimes the body was covered with fine cotton cloth, over which were placed finely woven blankets or ponchos of the wool of the vicuña or the alpaca, with designs in various colors.

The body and its wrappings were bound together by a net-work of stout cord of vegetable fibre; by a piece of strong cloth sewed together in the form of a closed sack, or in some localities by a casing of woven rushes.   The "mummy bundle" was surmounted by the so-called "false head," which was sewed to its upper surface.   The significance of this practice is unknown.   These false heads, many of which are present in the collection, were made of cloth and filled with different vegetable substances.   The face was represented in various ways:

Mummy Bundles   sometimes by a mask of wood or clay, but often the eyes, nose and mouth were made of wood, shell, gold or silver and fastened directly to the cloth by means of thread.   To the outside of the mummy bundle were often attached several of the prayer sticks or sepulchral tablets which are frequently found in considerable numbers in the sand about the grave.   These are either in the form of a cross wound with variously colored yarns, or a framework of split reeds, covered with cloth upon which rude designs are painted.   Favorite animals were sometimes buried with the dead as is shown by the mummified bodies of a dog and a parrot in the collection.

The manner of interment of the mummy bundle and its accompaniments differed in various localities.   In the coast region many of the

Huacas   mummies are found in little vaults, or "huacas," of adobes or flat stones roofed with sticks or canes, overlaid with mats or a layer of rushes, which prevented the earth covering from filling the grave.   These vaults usually contain from one to four bodies.

NATURALLY MUMMIFIED BODY

Copper Mine at Chuquicamata, Chile

10

Burials in stone towers or "chulpas" seem to have been confined chiefly to the Aymará Indians of the Callao, the great plateau of the Andes which includes the basin of Lake Titicaca and lies between the two maritime cordilleras and the eastern range, out of which rise the lofty volcanic peaks of Illimani and Sorata. In plan these chulpas are either circular or rectangular and are spoken of as round or square towers. A round burial tower is shown on page 2. Dr. von Tschudi found chulpas in the Department of Junin, which may have been built by Aymará *mitimaes*, or translated colonies. Describing the burial towers near Palca, E. G. Squier says:[1] "Primarily these chulpas consisted of a cist, or excavation, in the ground about four feet deep and three feet in diameter, walled up with rough stones. A rude arch of converging and overlapping stones, filled in or cemented together with clay, was raised over this cist, with an opening barely large enough to admit the body of a man, on a level with the surface of the ground, towards the east. Over this hollow cone was raised a solid mass of clay and stones, which, in the particular chulpa I am now describing as a type of the whole, was 16 feet high, rectangular in plan, 7½ feet face by 6 feet on the sides. The surface had been rough-cast with clay, and over this was a layer of finer and more tenacious clay or stucco, presenting a smooth and even surface."

Chulpas

One of the most remarkable specimens that the Department of Ethnology has acquired is a naturally mummified body which was found in an old copper mine at Chuquicamata, Province of Antofagasta, Chile, and which is illustrated on page 10. The condition of the body shows that the unfortunate miner was caught by a cave-in of the roof and partly crushed. The mummification seems to have been produced in part by the action of copper salts and not to have been altogether a desiccation due to the dryness of the region. The skin has not collapsed on the bones, as in the mummies found usually in the region, but the body and limbs preserve nearly their natural form and proportions, except for the crushing already mentioned. No analysis has yet been made of the tissues, so that it is too early to hazard any supposition as to the chemical changes which they have undergone. Mines in this neighborhood have been worked for an unknown length of time upon a peculiar deposit of atacamite, a

Natural Mummy

[1] Squier's Peru, p. 243.

hydrous chloride of copper, which is much prized on account of its easy reduction. The age of the mummy is unknown, but it is supposed to be pre-Columbian.

The story told by the objects found with the Peruvian dead is in part the story of ancient Peruvian life. The objects in the

Weapons and Implements
Peruvian collection in the hall, most of which have come from graves and mummy bundles, have been arranged so as to tell part of this story. For example we find with the bodies of men slings for throwing stones, stone-headed clubs and bolas (rounded stones joined by cords), showing the weapons and implements of war and the chase. With the mummy bundle of the woman have been found work-baskets, filled with threads and yarns of various colors, needles of thorn and copper, the implements used in weaving, such as spindles and shuttles, or the stones used in smoothing and polishing the outside of pottery vessels. Woman's work in ancient Peru is indicated by the presence or absence of objects familiar to us of the present day. Corn, beans and other foods were usually placed beside the body in the grave, together with vessels used in eating and drinking. These objects indicate not alone the belief of the people in a future world and the ne-

EAR OF CORN. FOUND WITH A MUMMY

cessity of sustaining the spirit in its journey thereto, but they also show that the people were well advanced in agriculture, and we are enabled to determine the kinds of plants cultivated and in many cases even the methods by which they were prepared for use.

Furthermore the objects found in the graves prove that in the working of copper, silver and gold the ancient Peruvians take high rank, and

CUP OF BEATEN GOLD AND STRING OF GOLD BEADS

13

show that the people knew how to exploit and treat the ores occurring in their land. Among copper implements there may be seen in the collection a great variety of spear points, club-heads, digging and planting implements, knives and axes. Tweezers are among the most familiar objects from the graves, and are often found suspended from the neck of a mummy by means of a cord.

Use of Copper

Some of the most notable of the gold objects are a cup beaten from a single piece, and ornamented in repoussé-work; human and animal figures, both solid and hollow, and beads and pins. The illustration on page 13 shows the gold cup and a string of large gold beads. In silver there are cups and vessels which, like the gold cup, are beaten from single pieces and are often ornamented with human or animal figures and other designs. Silver tweezers in many fanciful forms, pins and a variety of ornaments have been found in and with the mummy bundles. These objects prove that the makers were familiar with the processes of casting in moulds, beating and soldering. Many of the hollow figurines were made in three or more pieces and the parts soldered together.

Gold and Silver

Another remarkable class of objects is to be found on the right as one enters the hall. Here are many garments and pieces of cloth which were found wrapped around the dead or deposited in the graves. A glance at this part of the collection will show the ancient Peruvians had great skill in the art of weaving. Upon closer examination it will appear that they were familiar with most of the weaves known to modern people, from the finest gobelins to the coarsest cotton cloth. Many of the specimens cannot be excelled at the present time. The looms used were of the simplest description, consisting of two crosssticks, one at the top, and the other at the bottom. The warp threads were stretched from one to the other, while the woof or filling was passed over and under these by a shuttle. So the weaving of these most perfect fabrics may be said to have been by hand. In this respect they may stand in contrast to the modern machine methods. In addition to the excellence of weave Peruvian cloth is unique in decoration. The designs are woven in and consist of geometric figures and conventionalized representations of men, pumas, jaguars and various kinds of birds and fish. Some of the forms are illustrated on page 16. A part of the decorative effect is due to the regular repetition of the same design in different colors.

Cloth and Weaving

That the Peruvian should also take high rank as a potter will be gathered from even a superficial study of the collection of all forms of pottery at the west end of the hall, since many of the vessels show real beauty of outline and form and excellence in their painted decoration. These qualities seem the more remarkable when we consider that the

PIECES OF CLOTH FOUND WITH MUMMIES

makers had no knowledge of the potter's wheel and were unacquainted with the art of glazing. Some of the vessels were shaped by hand, but others show that they were formed by means of moulds. The body was moulded in two parts which were joined by being pressed together. Spout or handle, if desired, was then attached, and all irregularities in the junctures remedied by scraping and rub-

Pottery

POTTERY WATER-JAR WITH CORN DECORATION

17

bing. Moulds were often used in making many of the animal heads
and human figures that adorn these vessels. The decoration was put
on with paint, and, after firing, the vessels were polished by rubbing
with a smooth pebble.

In the absence of an aboriginal written language in Peru and on
account of the meagreness of the descriptions left by the first Europeans
who visited the country, it is fortunate for the student of Peruvian archæology that the potter often represented by the shape of his vessel or in its

POTTERY VESSEL WITH PAINTED DECORATION

decoration forms and customs which were familiar to him in his every-
day life.

Representations of the human figure are common. Some of these
show the manner of wearing the poncho and other articles of clothing.
Some have in the lobe of the ear the large cylindrical ear-ornaments
which led the Spaniards to nickname these people "Ore-
jones" — big ears. It would be impracticable, however, to   **Human
figures**
mention here more than a few of the subjects depicted. On
one vessel a man pursues and kills a deer with a spear; on another a
hunter is returning with the body of a deer thrown across his shoul-

ders. Some jars show the manner of catching fish by means of hook and line, while others portray men and women carrying water jars and other burdens by means of a strap passing around the forehead. Here we see a man with his favorite bird, evidently of the parrot family, perched upon his shoulder; there a dance in progress, with several of the figures playing upon musical instruments.

These potters were very fond of moulding their clay into animal forms, and they have left us more or less truthful representations of many of the species familiar to them. Their favorite models appear to
Animal figures   have been the puma, jaguar, monkey, llama, Guinea-pig, lizards, birds of the parrot family, the king vulture and a number of shells and vegetable forms. A complete list would include most of the animal and many of the vegetable forms of Peru.

Everywhere, except in the most elevated parts of the country, maize was not only the staple food of the people, but also was the source of their favorite intoxicating beverage,— *chicha*; hence it was but natural that they should so often represent the grain on their vessels. This
Chicha   was very simply and perfectly accomplished. A mould was made from an ear of corn and dried in the sun or fired. Into this clay was pressed; which on being removed would be a facsimile of the ear. This was joined to the jar while both were still in a plastic condition, after which the whole was fired and polished. A corn jar is represented in the illustration on page 17.

Although this guide relates chiefly to burials, it may not be out of place to call attention to some peculiarities of Peruvian skulls. The skulls of all races are of great scientific value, but those of Peru are of particular interest, because many of them bear the marks of surgical or
sacrificial operations. The Museum collection of Peruvian
Trephined skulls   skulls is so extensive that only a representative series is on exhibition. This contains many examples showing trephining, artificial deformation and pathological conditions, together with several normal Peruvian skulls for purposes of comparison.

In Peru, where stones from slings and wooden clubs with heads of stone and copper were the common offensive weapons, complex fractures of the skull with depression of its bony plates must have been common. There seems no reason to doubt that trephining was resorted to as a means of relief in such fractures, and that sometimes cures were effected by this treatment. It is also probable that the operation in many

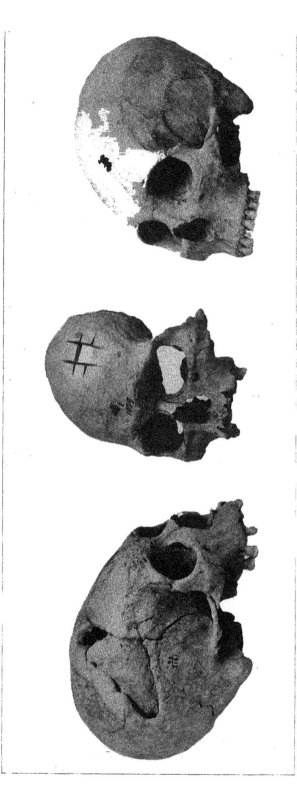

TREPHINED SKULLS FROM ANCIENT PERUVIAN GRAVES

cases was a part of some religious ceremonial, since some of the trephined skulls in the collection show distinct orientation of the wound and present no indication of lesion. Implements of copper and bronze and knives of stone and obsidian must have been employed in the operation, which was performed with skill.

Artificial deformation of the head was extensively practised in ancient Peru and was accomplished by means of ligatures applied in infancy. The form taken by the head was determined by the manner in which these bindings were applied. The pathological skulls show the ravages of disease in the bones of the cranium.

THE QUIPU.

The Quipu is a fringe consisting of a main cord with other cords of various colors hanging from it. In the fringe knots of different kinds were tied. The ancient Peruvians, having no written language, made use of the quipu to keep their accounts and possibly to record historic incidents. By the color of the cord, the kind of knot, the distance of the knots from the main cord and from each other, many facts could be recorded and preserved. The maker of a quipu had a system which was to a great extent arbitrary, and which had to be explained when the quipu was placed in the keeping of another.

COCA CHEWING.

The coca plant (*Erythroxylon coca*, Lam.) grows wild in the mountainous regions of Peru and Bolivia and was cultivated before the Conquest, as it is to-day, in districts from 2,000 to 5,000 feet above the sea. It is valued for its stimulating narcotic properties, and the present Indians will often carry heavy burdens for several days without food, if furnished with a plentiful supply of coca. The leaves are gathered and dried in the sun and then chewed mixed with unslacked lime in the same way the betel is used by the East Indians. A bag of coca teaves is almost always found with a mummy. The leaves of this plant, together with the cloth bags in which they were carried and the gourd flasks containing lime may be seen in the collection.

MISCELLANEOUS EXHIBITS.

This gallery contains many exhibits, some of them very important, of which no special mention has been made, since it is believed that the case labels and the guide leaflets attached to the cases will furnish the desired information to the student and visitor. Among these may be mentioned the collection from the West Indies, the musical instruments of the Incas, the case containing a great variety of animal forms in pottery; collections of feather-work from Peru, Bolivia, Paraguay and Brazil, and the collection from Columbia consisting of many objects in pottery, stone and shell.

BIBLIOGRAPHY.

The-following books will be found useful to those who wish to study South American Archæology and they may be consulted on application to the librarian of the Museum.

VEGA, GARCILASSO DE LA. The Royal Commentaries of Peru. Ed. Rycaut, London, 1688.

CIEZA DE LEON, PEDRO DE. Chronicle of Peru, Part I. (Hakluyt Society.) London, 1864.

ANDAGOYA, PASCUAL DE. Narrative of the Proceedings of Pedrarias Davila. (Hakluyt Society.) London, 1865.

PRESCOTT, WILLIAM H. History of the Conquest of Peru. London, 1847.

HERNDON AND GIBBON. Exploration of the Valley of the Amazon. Washington, 1853.

PREZIER, MONSIEUR. A Voyage to the South Sea and Along the Coast of Chili and Peru. London, 1717.

WHYMPER, EDWARD. Travels Amongst the Great Andes of the Equator. New York, 1892.

HUTCHINSON, THOMAS, JR. Two Years in Peru, with Explorations of its Antiquities. London, 1873.

SQUIER, E. G. Incidents of Travels and Explorations in the Land of the Incas. New York, 1877.

WIENER, CHARLES. Perou et Bolivie. Paris, 1880.

REISS AND STUEBEL. The Necropolis of Ancon. Ed. Keane, New York, 1887.

BÆSSLER, ARTHUR. Ancient Peruvian Art. New York, 1903.

STUEBEL AND UHLE. Die Ruinenstätte von Tiahuanaco. Breslau, 1892.

HOLMES, WILLIAM H. Textile Art in its Relation to the Development of Form and Ornament. Annual Report Bureau Amer. Ethnology, 1884–85.

STUEBEL, REISS AND KOPPEL. Südamerikanische Völker. (Alte Zeit.) Berlin, 1889.

NORTHEAST QUARTER OF FOYER

Showing five of the Marble Busts of Pioneers of American Science.

PIONEERS

OF

AMERICAN SCIENCE

AN ACCOUNT OF THE EXERCISES HELD

AND THE ADDRESSES DELIVERED

AT THE

AMERICAN MUSEUM OF NATURAL HISTORY

DECEMBER 29, 1906

PIONEERS OF AMERICAN SCIENCE.

Saturday, December 29, 1906, was notable in the annals of the Museum, since at 3 o'clock of that day there were held in the large auditorium the ceremonies attending the unveiling of the marble busts which have been installed in the foyer, representing ten of the men who have been foremost in the advancement of science in America. The auditorium was crowded to its full capacity with Members of the Museum and their friends and visiting scientists, and Professor H. F. Osborn, Second Vice-President of the Museum, presided in the absence, due to illness, of Mr. J. Pierpont Morgan, First Vice-President. The exercises began with the singing of the national anthem "America," after which Dr. Hermon C. Bumpus, Director of the Museum, acting for Mr. Morris K. Jesup, addressed the Trustees as follows:

"Thirty-six years ago several men of this city organized to perform three closely-related functions:

1, To establish and maintain a museum of natural history;

2, To encourage and develop the study of the natural sciences;

3, To advance the general knowledge of kindred subjects.

"Of this company, Mr. Morgan was one; a second (Mr. Choate) is he who will receive on behalf of the Honorable Board of Trustees the splendid gift that has brought this congregation of scientists together, and a third is he who for more than twenty-five years, as President, has devoted his time, his thought, his energy, his influence, his means, his health, not for the mere naked fulfilment of the terms of the Articles of Incorporation, but for the up-building of an institution that would excite civic pride, for the molding of forces that would result in educational power, for the combination of material that would develop character and for the general exploration of the secrets of Nature, be they hidden in the remote regions of Siberia, in the unknown land and waters of the North, in the ancient monuments of the South or the outcropping foundations of the continent in the West.

"To what extent the sympathetic union of these three men with other earnest workers in a common cause has been successful in the establishment and maintenance of a museum of natural history, none are better able to judge than the members of the various scientific and educational organizations,— the guests that honor the Museum by their presence this afternoon. Many of our guests today are frequent students of the Museum's collections, frequent readers of the Museum's publications and frequent users of the Museum's library.

"But it is the effort to fulfill the terms of the second paragraph of the Articles of Incorporation — the paragraph that refers to the encouragement and development of the Study of the Natural Sciences, that provides for the aggressive invasion of the unknown and for the encouragement of those who enter the unknown for search and research — it is the effort to fulfill these terms that has characterized the administration of this institution and made it something different from a mere museum.

"The prime incentive for the pursuit of science is doubtless indiscoverably hidden among those forces that direct the growth and activities of the human body, but the strongest secondary incentive is appreciation — commendation. An institution that is pledged to the encouragement and development of the natural sciences ought certainly to appreciate and commend those who have conspicuously devoted themselves to the pursuit of science. It is in response to this feeling of obligation and with the hope that such recognition at this time might act as a helpful incentive to those attending the important scientific meetings now being held in New York, that these exercises have been arranged.

"In an adjoining hall, as we leave this auditorium, we shall find unveiled ten portraits of the pioneers of American science, the work of one of America's leading sculptors, Mr. William Couper.

"It is this series of busts that I have the honor, on behalf of Mr. Morris K. Jesup, to present to the Trustees for permanent exhibition in this Museum, and with it may I convey the desire that they may serve as a token of the donor's esteem for all who have devoted and are devoting themselves to the development of science, and also that all entering this institution may feel that the study of the natural sciences is encouraged and developed, not by immediate results alone, but also in the proper recognition of those who have unselfishly labored for its advancement."

In accepting the gift, Hon. Joseph H. Choate said in behalf of the Board of Trustees:

"As you have already heard, it is only in my accidental capacity of survivor that I have the honor of appearing here today to receive this gift. Mr. Morgan could not come, Mr. Jesup could not accept his own gift, — he knows how much better and more delightful it is to give than to receive, — and so I stand here for a moment on behalf of my fellow trustees, to receive this splendid donation.

"If this were Mr. Jesup's only gift to the Museum, it ought to place him among the immortals. To place in our vestibule, at the entrance of these halls of science, the busts of these great pioneers and masters, to place them here so that the future generations of New York and of America may become familiar with their features would be in itself a very great and valuable gift. Ever since the foundation of this Museum thirty-seven years ago, he has been enriching and endowing it with wonderful gifts.

"Most of you are perfectly familiar with the chief of these, — the Jesup Collection of Woods, containing the wood of every tree known to be existing in North America, a perfectly unique collection which cannot anywhere be repeated; the collections that were brought by the Jesup North Pacific Expedition from the shores of British Columbia, Alaska and Asia are also unique in their way; and in the Hall of Vertebrate Palæontology a large portion of that wonderful exhibit is from his generous hands. Even now, today, he is fitting out for our benefit an expedition for the exploration of fossils in northern Egypt, and I am sure that when Professor Osborn, who is to head the expedition, returns next spring, he will come 'bearing his sheaves with him,' in the form of the fossil remains of the ancestral elephant, which he will find somewhere between the mouth of the Nile and the Nubian Desert — exactly where I cannot tell, but he, at this moment, with his prophetic vision could put his finger upon the very spot.

"This Museum, if you will notice the date, was born in the Dark Ages of the City of New York — in 1869 — when the public enemy was in possession of the city and of its treasury. It was a gloomy day for the foundation of such an institution. I believe it was about that time that one board of public officials, catching a strange ray of light for that dark time, had employed the celebrated Dr. Hawkins to prepare models of the vast fossil mammals for exhibition to the people. They gave him a house in Central Park, where he set to work on that great study. By and by, there came in another set of public officials who were as antediluvian as the fossils themselves, and they broke his models all up and sent the doctor on his way not rejoicing at all.

"We never dared in those days to hope or expect help from the

City for such an institution as this, but light soon dawned upon us, and gradually year by year the City Officials and the people of the City began to find that this was a great educational institution maintained for the benefit of the people. But it was hard struggling in those early days. Despair followed anxiety, and our Trustees knew not which way to turn. But when Mr. Jesup, twenty-six years ago, took the presidency of this body, he found that certain fossils still lingered in the Board of Trustees, and actually breathed into us the breath of life. He has kept us alive ever since, and every year this Museum has exhibited new energy and success, and more than three fourths of it is due to his generous gifts and his inspiring presence.

"He has given us something far better and grander than material assistance, liberal as he has been with that. He has given us twenty-six of the best years of his life, devoted with untiring generosity to the success of this enterprise. It is also to be remembered to the great credit of Mr. Jesup that it was during his administration that the Museun took one significant step forward, which we had long been hesitating to do — I mean the opening of the Museum to the public on Sundays. It was the best advance we ever made. We found that we could do our duty to the church in the morning and come here with equally reverent minds in the afternoon to study these collections and look through Nature up to Nature's God, and the people found that out too.

"And so with grateful hearts the Trustees accept this last and noble offering, and I am sure you will all with one voice join with me in saying — God bless the noble donor."

After Mr. Choate's address brief memorials of the men of science whose portraits have been selected for the foyer, were delivered in accordance with the following program:

BENJAMIN FRANKLIN, by Dr. S. Weir Mitchell of Philadelphia;
ALEXANDER VON HUMBOLDT, by His Excellency, Baron Speck von Stern-burg, German Ambassador. (Read by Major T. von Körner, Military Attaché of the Embassy);
JOHN JAMES AUDUBON, by Dr. C. Hart Merriam, Chief, U. S. Biological Survey, Washington, D. C.;
JOHN TORREY, by Dr. Nathaniel L. Britton, Director in chief, New York Botanical Garden, New York City;
JOSEPH HENRY, by Dr. Robert S. Woodward, President, Carnegie Institution, Washington, D. C.;
LOUIS AGASSIZ. Letters were read from the Rev. Edward Everett Hale, an

intimate personal friend of Professor Agassiz, and Professor F. W. Putnam, of Harvard University, and remarks were made by Professor Addisson E. Verrill of Yale University and Dr. C. D. Walcott, Director of the U. S. Geological Survey;

JAMES DWIGHT DANA, by Dr. Arthur Twining Hadley, President, Yale University, New Haven, Conn.;

SPENCER FULLERTON BAIRD, by Dr. Hugh M. Smith, Deputy Commissioner, Bureau of Fisheries, Washington, D. C.;

JOSEPH LEIDY, by Professor William Keith Brooks, Johns Hopkins University, Baltimore, Md.;

EDWARD DRINKER COPE, by Dr. Henry Fairfield Osborn, Curator, Department of Vertebrate Palæontology, American Museum of Natural History.

The addresses as delivered were as follows:

BENJAMIN FRANKLIN.

BY S. WEIR MITCHELL.

We are here, as I understand, to unveil memorial busts of Americans distinguished in science, and I am honored by the privilege of speaking of Benjamin Franklin. This man, the father of American Science, was possessed of mental gifts unequalled in his day. Even yet he holds the highest place in the intellectual peerage of a land, where, in his time, men had few interests which were not material or political. But no man entirely escapes the despotic influences of his period. Thus in every life there are unfulfilled possibilities, and so it was that, paraphrasing Goldsmith, we may say that Franklin to country gave up what was meant for mankind, when with deep regret he resigned in middle life all hope of whole-souled devotion to science. When most productive, his scientific fertility was the more remarkable because of the other forms of dutiful activity which, in a life that knew no rest, left small leisure for those hours of quiet thought without which science is unfruitful of result.

There is a Hall of Fame not built by the hand of man. It is the memory of mankind. In many of its galleries this man's bust could with justice be placed. Diplomacy would claim him as of her greatest. For him would be the laurel of administrative wisdom. Among statesmen he would be welcomed. Who of the masters of English prose shall in that hall of fame be more secure of grateful remembrance, and who more certain of a place among men of science?

As an investigator of Nature and of Nature's laws he is materially represented here by right of eminent achievement. Let us as men of science feel proud that Franklin's fame as a philosopher did much to win for Franklin the diplomatist such useful consideration and respect as led to final success.

Many of those you honor today had moral and temperamental peculiarities which more or less influenced their lives and are common to men of science. Most of them cared little about making money; still less about keeping it. Franklin on the contrary dreaded poverty; was careful in business, made fruitful investments and died rich; nevertheless like the typical man of science he refused to make money out of his discoveries, or to protect his inventions by patents. In him the man of science, unselfish, free from money greed, seemed to exist apart from all those other men who went to the making of the many-minded Franklin. In another way he was singularly unlike such typical men of science as Henry, in physics, and Leidy, in natural history. When Franklin made a discovery, his next thought was as to what practical use it could be put. If he made some novel observation of nature, he asked himself at once how he could make it serve his fellow men. The great reapers of the harvest of truth commonly leave the inventor to make practical use of their unregarded thought.

Leaving the wide land to do justice to Franklin, the model citizen and great diplomatist, here we crown with the assured verdict of posterity Franklin, the man of pure science. Here we welcome him to this goodly fellowship of those who communed with Nature and read the secrets of the Almighty Maker.

ALEXANDER VON HUMBOLDT.

Mr. President, Gentlemen:

His Excellency, the German Ambassador, whom heavy official duties retain at Washington, has requested me to represent him on this occasion and to express to you his hearty congratulations on this event on which through Mr. Jesup's generous munificence this commemorative tribute is paid to the world's great masters of science, a day on which this magnificent museum of natural history has received a donation which will awake a solemn sense of reverence and make this abode

ALEXANDER VON HUMBOLDT

Born, Berlin, September 14, 1769
Died, Berlin, May 6, 1859

Geographer, traveler, philosopher

Described the surface features and the geological structure of many lands

henceforth a temple of devotion to the founders and promoters of natural science.

Whoever honors the memory of great men, honors science and honors himself, and so the Ambassador has asked me to convey to the generous donor, the tireless promoter of science, Mr. Jesup, the expression of his sincerest admiration and of his heartfelt thanks for the honor which will be conferred also upon the great German scholar, Alexander von Humboldt.

In this immortal man, whose bust you have gathered to unveil, the world reveres its greatest master since the days of Aristotle. His genius covered all that man has ever thought, done and observed in nature. There is no branch of human knowledge into which his mind did not penetrate. His "Cosmos," that marvellous monument of meditation and research, is a new book of Genesis in which the Universe mirrors itself in all its vastness and minuteness "from the nebulæ of the stars" — to use his own words — "to the geographical distribution of mosses on granite rocks."

By his wonderful talent of research, by his almost superhuman power to divine eternal laws, this great interpreter of science taught mankind how to read the book of nature, how to understand its great mysteries. The series of sciences, originated by this mighty genius is, as well as the other manifold branches of science developed by him, sufficiently known to all.

In all his investigations his ultimate aim was to bring theory into practical relation with life. Thus he not only elevated the standard of culture of the whole world by many steps, but he also became from a practical point of view the benefactor of mankind in many branches of common life,— as trade, commerce, navigation.

He taught us how to conceive the beauty and sublimity of nature in its every form and motion. His studies are not a matter merely of memory and of dry meditation, to him Nature was rather the inexhaustible source of pure and deep enjoyment, by which the heart is purified and ennobled and men are brought nearer to perfection.

It is not necessary to give you a more detailed picture of his life. All this is so well known and so dear to the whole learned world of America; for never has a foreign scholar been more honored in this country than Alexander von Humboldt. To realize this we need only recall the celebrations which took place in his memory throughout all

America both at the time of his death and on the occasion of the centennial anniversary of his birth.

Humboldt devoted five years of his life to scientific investigations in South and Central America, in Mexico and in Cuba. He ascertained the course of the greatest rivers; he climbed the summits of mountains where man's foot had never trod before; he studied vegetation, astronomical and meteorological phenomena, gathered specimens of all natural products and a great deal of historical information about the early population of these parts of the New World. It was he that drew the first accurate maps of these regions. With almost prophetic forecast of the needs of generations to come, he examined the Isthmus of Panama and considered carefully the possibility of establishing there an interoceanic waterway.

It is well known how great an interest Alexander von Humboldt took in the United States. Indeed, so strongly was he attracted by the problems of the new-born Republic that putting aside even his habitual scientific occupations, he devoted himself entirely for some months to the study of the American people and the institutions of this country.

Finally, the great scientist, he whom people call the scientific discoverer of America, returned to his country, carrying with him a vast store of intellectual and material treasures of science. So abundant were the results, reaped from his expeditions, that he needed the coöperation of the best scholars of his time to compile that great mass of material, and to place it in proper shape and form.

Throughout his long and industrious life, Alexander von Humboldt ever retained his love for and devotion to the country where his great field of labor lay, and for its people with whom he always felt closely connected by his love for freedom in thought and for liberty. It is a well-known fact that in his later days of all the foreigners, who knocked at his door, no one was more heartily welcomed than the American citizen.

The benefits of his investigations in America returned to that country in the course of time. No wonder that her people recognize him as their benefactor. Another great man, whose monument will be unveiled today, and most deservedly placed beside the one of Alexander von Humboldt, Louis Agassiz, says of him:

"To what degree we Americans are indebted to von Humboldt, no

10

JOHN JAMES AUDUBON

Born, New Orleans, May 4, 1780
Died, New York, January 27, 1851

Ornithologist, artist

Chief work, "The Birds of America," illustrated with 1065 life-size colored plates

one knows who is not familiar with the history of learning and education in this country. All the fundamental facts of popular education in physical science beyond the merest elementary instruction, we owe to him." At another place he says, "Let us rejoice together that Humboldt's name will permanently be connected with education and learning in this country, for the prospects and institutions of which he felt so deep and so affectionate a sympathy."

Of all the tributes that have been paid to Alexander von Humboldt the most lasting and most fitting has now found its expression in this building. For here, in this magnificent American Museum of Natural History the ideal aim of all his theories is realized most perfectly: to cultivate the love of Nature, and thus to ennoble man and beautify his life.

Gentlemen, permit me to thank you for the honor you have done me today, and to express the hope that this splendid building may become a shrine of pilgrimage of scientists and students also of the Old World, helping to bind the nations closer together.   ·

Baron SPECK VON STERNBURG,
represented by Major THEODORE KÖRNER.

JOHN JAMES AUDUBON.

BY C. HART MERRIAM.

Of the naturalists of America no one stands out in more picturesque relief than Audubon, and no name is dearer than his to the hearts of the American people. Born at an opportune time, Audubon undertook and accomplished one of the most gigantic tasks that has ever fallen to the lot of one man to perform. Although for years diverted from the path Nature intended him to follow, and tortured by half-hearted attempts at a commercial life, against which his restive spirit rebelled, he finally broke away from bondage and devoted the remainder of his days to the grand work that has made his memory immortal.

His principal contributions to Science are his magnificent series of illustrated volumes on the Birds [1] and Quadrupeds [2] of North America, his Synopsis of Birds,[3] and the Journals [4] of his expeditions to Labrador and to the Missouri and Yellowstone Rivers.

The preparation and publication of his elephant folio atlases of life-size colored plates of birds, begun in 1827 and completed in 1838, with the accompanying volumes of text (the "Ornithological Biography," 1831-1839), was a colossal task. But no sooner was it accomplished than an equally sumptuous work on the mammals was undertaken, and, with the assistance of Bachman, likewise carried to a successful termination. For more than three-quarters of a century the splendid paintings which adorn these works, and which for spirit and vigor are still unsurpassed, have been the admiration of the world.

In addition to his more pretentious works, Audubon wrote a number of minor articles and papers and left a series of "Journals," since published by his grand-daughter, Miss Maria R. Audubon. The Journals are full to overflowing with observations of value to the naturalist and, along with the entertaining "Episodes," throw a flood of light on contemporary customs and events. Incidentally, they are by no means to be lost sight of by the historian.

In searching for material for his books Audubon traveled thousands of miles afoot in various parts of the eastern states, from Maine to Louisiana; he also visited Texas, Florida and Canada; crossed the ocean several times, and conducted expeditions to far-away Labrador and the then remote Missouri and Yellowstone Rivers. When we remember the limited facilities for travel in his day, the scarcity of railroads, steamboats and other conveniences, we are better prepared to appreciate the zeal, determination and energy necessary to accomplish his self-imposed task.

That it was possible for one man to do so much excellent field work, to write so many meritorious volumes and to paint such a multitude

[1] *The Birds of America*, 4 atlases, double elephant folio colored plates. London, 1827-1838; Ornithological Biography, an account of the habits of the birds of the United States. 5 vols. Royal 8vo, Edinburgh, 1831-1839.

[2] *The Quadrupeds of North America* by John James Audubon and Rev. John Bachman. 3 vols. Royal 8vo text, and elephant folio atlas of colored plates. New York, 1846-1854.

[3] *Synopsis of Birds of North America*. Edinburgh & London, 1839.

[4] *Audubon and his Journals* by Maria R. Audubon. 2 vols. 8vo. New York, 1897.

JOHN TORREY

Born, New York, August 15, 1796
Died, New York, March 10, 1873

Botanist, chemist

One of the founders of botanical science in the United States

of remarkable pictures must be attributed in no small part to his rare physical strength, for do not intellectual and physical vigor usually go hand-in-hand and beget power of achievement? Audubon was noted for these qualities. As a worker he was rapid, absorbed and ardent; he began at daylight and labored continuously till night, averaging fourteen hours a day, allowing, it is said, only four hours for sleep.

In American ornithology, in which he holds so illustrious a place, it was not his privilege to be in the strict sense a pioneer, for before him were Vieillot, Wilson and Bonaparte; and contemporaneous with him were Richardson, Nuttall, Maximilian Prince of Wied and a score of lesser and younger lights some of whom were destined to shine in the near future.

Audubon was no closet naturalist — the technicalities of the profession he left to others — but as a field naturalist he was at his best and had few equals. He was a born woodsman, a lover of wild nature in the fullest sense, a keen observer and an accurate recorder. In addition he possessed the rare gift of instilling into his writings the freshness of nature and the vivacity and enthusiasm of his own personality.

His influence was not confined to devotees of the natural sciences, for in his writings and paintings, and in his personal contact with men of affairs both in this country and abroad, he exhaled the freshness, the vigor, the spirit of freedom and progress of America, and who shall attempt to measure the value of this influence to our young republic?

Audubon's preëminence is due not alone to his skill as a painter of birds and mammals, or to the magnitude of his contributions to science, but also to the charm and genius of his personality, a personality that profoundly impressed his contemporaries, and which, by means of his biographies and journals, it is still our privilege to enjoy. His was a type now rarely met, combining the grace and culture of the Frenchman with the candor, patience and earnestness of purpose of the American. There was about him a certain poetic picturesqueness and a rare charm of manner that drew people to him and enlisted them in his work. His friend, Dr. Bachman of Charleston, tells us that it was considered a privilege to give to Audubon what no one else could buy. His personal qualities and characteristics appear in some of his minor papers, notably the essays entitled "Episodes." These serve to reveal, perhaps better than his more formal writings, the keenness of his insight, the kindness of his heart, the poetry of his nature, the power of his imagination and the vigor and versatility of his intellect.

JOHN TORREY.

By Nathaniel L. Britton.

As a pioneer of American botany, John Torrey naturally finds a place among the men whose works we gladly celebrate today in this grand institution developed in the city where he was born; where he resided the greater part of his life, and where he died. Today's recognition of Torrey as a master of botanical science, is therefore peculiarly appropriate in New York, where he is already commemorated by the society which bears his name; by the professorship in Columbia University named in his honor, and by his botanical collections and library deposited by Columbia University at the New York Botanical Garden.

Dr. Torrey was born August 15, 1796, and died March 10, 1873, nearly thirty-four years ago; the pleasure of his personal acquaintance is therefore known to but few persons now living. We have abundant evidence, however, that he was honored and beloved to a degree experienced by but few; righteousness was instinctive in him; aid to others was his pleasure; he was tolerant and progressive, and his genial presence was a delight to his associates.

He was educated for the profession of medicine, graduating from the College of Physicians and Surgeons in 1818, but he soon abandoned it and in 1824 became professor of chemistry at West Point; after three years service there, he was elected professor of chemistry and botany in the College of Physicians and Surgeons, a position which he held for nearly thirty years, during part of this period lecturing on chemistry also at Princeton: he was also United States assayer in New York from 1854 until his death.

Dr. Torrey's attention was directed to botany during his youthful association with Professor Amos Eaton, and his interest in that science was subsequently stimulated during his medical studies by the lectures of Professor David Hosack. It early became his favorite study, and, notwithstanding his noteworthy services to chemistry, his fame rests on his botanical researches, although they were accomplished during his hours of rest and recreation, — largely during the night.

His botanical publications began in 1819 with "A Catalogue of Plants Growing Spontaneously within Thirty Miles of the City of New York," published by the Lyceum of Natural History, now the New York Acad-

JOSEPH HENRY

Born, Albany, N. Y., December 17, 1797
Died, Washington, D. C., May 13, 1878

Physicist

Noted for his investigations in electromagnetism
First secretary of the Smithsonian Institution

emy of Sciences, and were completed the year after his death in the "Phanerogamia of Pacific North America," in Vol. 17 of the Report of the United States Exploring Expedition. His contributions to botany include more than forty titles, many of them volumes requiring years of patient study; they throw a flood of light on the plants of North America, and form a grand contribution to knowledge. His collections, on which these researches are based, were annotated and arranged by him with scrupulous care and exactness, and are treasured as among the most important of all scientific material in America.

JOSEPH HENRY.

By Robert S. Woodward.

This time, one hundred years ago, Joseph Henry, whose name and fame we honor today, was a lad seven years of age. He was born at Albany, New York, of Scotch parentage, his grand parents on both sides having come from Scotland in the same ship to the Colony of New York, in 1775.

Doubtless he had himself in mind when in his mature years he affirmed that "The future character of a child, and that of a man also, is in most cases formed probably before the age of seven years." At any rate, he found himself early, for at the age of sixteen he had determined to devote his life to the acquisition of knowledge. Thus he became, in turn, student; teacher; civil engineer in the service of his native State; professor of mathematics and natural philosophy in the Albany Academy; professor of natural philosophy in the College of New Jersey — now Princeton University — and a pioneer investigator and discoverer of the first order before he was thirty-three years of age.

His inventions and discoveries in electromagnetism especially are of prime importance. They include the inventions of the electromagnetic telegraph and the electromagnetic engine and the discovery of many of the recondite facts and principles of electromagnetic science.

From the age of thirty-three, when he took up the work of his professorship at Princeton, till the age of forty-seven, when he was called to the post of Secretary of the Smithsonian Institution, he pursued his original investigations with untiring zeal and with consummate experimental skill and philosophic insight. It was during this period that Henry and Faraday laid the foundations for the recent wonderful developments of electromagnetic science. The breadth as well as the depth of Henry's learning is indicated by the fact that he found time during this busy period for excursions and for lectures in the fields of architecture, astronomy, chemistry, geology, meteorology, and mineralogy in addition to his lectures and researches in physics.

He was a man rich in experience and ripe in knowledge when, in 1846, he assumed the administrative duties implied by the bequest of James Smithson, "To found at Washington, under the name of the Smithsonian Institution an Establishment for the increase and diffusion of knowledge among men." Thenceforth, for thirty-two years, until his death in 1878, he devoted his life to the public service, not alone of our own country, but of the entire civilized world. In this work he manifested the same creative capacity that had distinguished his earlier career in the domain of natural philosophy. He became an organizer and a leader of men. To his wise foresight we owe not only the beneficent achievements of the Smithsonian Institution itself, but also, in large degree, the correspondingly beneficent achievements of the Naval Observatory, the Coast and Geodetic Survey, the Weather Bureau, the Geological Survey, the Bureau of Fisheries and the Bureau of American Ethnology; for to Henry, more than to any other man, must be attributed the rise and the growth in America of the present public appreciation of the scientific work carried on by governmental aid.

We may lament, with John Tyndall, that so brilliant an investigator and discoverer as Henry should have been sacrificed to become so able an administrator. And American devotees to mathematico-physical science may be pardoned for entertaining an elegiac regret that Henry as a pioneer in the fields of electromagnetism did not have the aid of a penetrating mathematical genius, as Faraday had his Maxwell. But posterity, just in its estimates towards all the world, will recognize in Henry, as we have recognized in our earlier hero, Benjamin Franklin, a many-sided man — a profound student of Nature; a teacher whose moral and intellectual presence pointed straight to the goal of truth; an inventor who dedicated his inventions immediately to the public good; a discoverer of the permanent laws which reign in the Sphinx-like realm of physical phenomena; an administrator and organizer of large enterprises which have yielded a rich fruitage for the enlighten-

LOUIS AGASSIZ

Born, Motier, Switzerland, May 28, 1807
Died, Cambridge, Mass., December 14. 1873

Zoölogist, ichthyologist

Celebrated lecturer and writer on natural history in general

ment and for the melioration of mankind; a leader of men devoted to the progress of science; a patriot, friend and counsellor of Abraham Lincoln in the darker days of the Republic — in short, an exemplar for his race, a man whose purity and nobility are here fitly symbolized in enduring marble for our instruction and guidance and for the instruction and the guidance of our successors in the centuries to come.

LOUIS AGASSIZ.

A LETTER FROM EDWARD E. HALE.

Read by ADDISON E. VERRILL, who added interesting personal reminiscences of Agassiz.

Washington, D. C., December 8, 1906.

I think that the first time when I ever saw Agassiz was at one of his own lectures early in his American life. This was a description of his ascent of the Jungfrau. I think it was wholly extempore, and, though he was new in his knowledge of English, it was idiomatic and thoroughly intelligible. At the end, as he described the last climb, hand and foot, by which, as it seems, men come to the little triangular plane, only three feet across, which makes the summit, he quickened our enthusiasm by describing the physical struggle by which he lifted himself so that he could stand on this little three-foot table. He said, "one by one we stood there, and looked down into Swisserland." He bowed and retired.

I know I said at once that Mr. Lowell, of our Lowell Institute, who had "imported Agassiz," (that is James Lowell's phrase) might have said before the audience left the hall, "You will see, ladies and gentlemen, that we are able to present to you the finest specimen yet discovered of the genus *homo* of the species *intelligens*."

And looking back half a century, on those very first years of his life in America, I think it is fair to say that wherever he went he awakened that sort of personal enthusiasm. And he went everywhere. He was made a professor in Harvard College in 1848. But he never thought of confining himself to any conventional theory of a college professor's work. He was not in the least afraid of making science popular. He

17

flung himself into any and every enterprise by which he could quicken the life of the common schools, and in forty different ways he created a new class of men and women. Naturalists showed themselves on the right hand and on the left. I have seen him address an audience of five hundred people, not twenty of whom when they entered the hall thought they had anything to do with the study of Nature. And when after his address they left the hall, all of the five hundred were determined to keep their eyes open and to study Nature as she is. From that year 1848, you may trace a steady advance in Nature Study in the New England schools.

That is to say, that his distinction is that of an educator quite as much as it is that of a naturalist. In 1888, Lowell said, in his quatermillennial address at Harvard College, that the College had trained no great educator, "for we imported Agassiz." A great educator he truly was.

When Agassiz was appointed Professor he was forty-one years old. In my first personal conversation with him he told me a story which may not have got into print, of his own physical strength. He spoke as if it were then an old experience to him. Whether he were twenty-five or thirty-five when it happened, it shows how admirable was his training and his physical constitution. He had been with a party of friends somewhere in eastern Switzerland. They were travelling in their carriages; he was on foot. They parted with the understanding that they were to meet in the Tyrol, at the city of Innsbruck. Accordingly the next morning, Agassiz rose early and started through the mountains by this valley and that, as the compass might direct or his previous knowledge of the region. He did not mean to stop for study and they did not. But he had no special plan as to which hamlet or cottage should cover him at night. Before sundown he came in sight of a larger town than he expected to see, in the distance, and calling a mountaineer, he asked him what that place was. The man said it was Innsbruck. Agassiz said that that could not be so. The man replied with a jeer that he had lived there twenty years, and had always been told that that was the name of the place, but he supposed Agassiz knew better than he did. Accordingly Agassiz determined that he would sleep there and did so. The distance was somewhere near seventy miles. I know it gave me the impression of a walk through the valley passes at the rate of four miles an hour, for sixteen or seventeen hours.

JAMES DWIGHT DANA

Born, Utica, N. Y., February 12, 1813
Died, New Haven, Conn., April 14, 1895

Geologist, mineralogist, zoölogist

One of the principal founders of geological science in the United States

In later life Agassiz made to us some prophecies in which we may trace his enjoyment of the finest physical health and strength. Health and strength indeed belonged to everything which he said and did.

Among other things he said, twenty-five years ago, that the last years of our century, — the twentieth, would see a population of a hundred million of people in the valleys of the upper Amazon. I like to keep in memory this brave prophecy, because I am sure it will come true.

FROM A LETTER FROM PROFESSOR F. W. PUTNAM.

Read by CHARLES D. WALCOTT, who gave also his own token of appreciation of Agassiz.

"It is a real grief to me that I cannot take this opportunity to offer tribute to my beloved and honored teacher, — Louis Agassiz. What a pleasure it would be to me to say a few words of appreciation of that great and good man. Not alone to speak of his scientific achievements, which are known the world over, but, from my intimate association with the great naturalist, to tell of all he did, fifty years ago, for the advancement and encouragement of the study of natural history; to picture his inspiring method of teaching; and to dwell on his goodness of heart, his genial magnetic personality and his wonderful power of winning the life-long devotion of his students."

JAMES DWIGHT DANA.

BY ARTHUR T. HADLEY.

It was my privilege to know James Dwight Dana intimately during my early years. To boyhood's imagination his figure typified the man of science; his life personified the spirit of scientific discovery. Wider acquaintance with the world has not in any way dimmed the brightness of that early impression.

The services of the geologist are today recognized by every one, and sought by all who can afford them. If he would make a voyage of exploration and discovery, the resources of the world of finance are placed at his disposal. No such aids were given two generations ago. In Dana's journeyings he had to surmount hardship and peril, and to meet the coldness of those who knew not the value of the quest which he pursued. He and his contemporaries were like the knights errant of chivalry, devoting their lives to an ideal. They were men of faith, who combined the spirit of the missionary and the inspiration of the poet with the clear vision of the observer.

The largeness of Dana's work was commensurate with the largeness of his inspiration. It fell to his lot not only to fill out many pages of the record of the building of the world, as written in the fossil life of America, but to show in important ways the methods by which that building was accomplished. His creative brain never rested content with mere description of facts. He had the more distinctively modern impulse to reconstruct the process by which those facts were brought to pass. From his observations of coral islands in the various stages of their growth he deduced a geologic principle of world-wide importance. It is this characteristic which makes the great modern German school of geologists headed by Suess look to Dana as their precursor, more than to any other man of his generation.

He was not content with the work of discovery alone. The teaching spirit was strong within him. The pioneers in science needed editors and expositors who should make their results known. In each of these capacities Dana's achievements were phenomenal. Of his work as an editor he has left the files of the American Journal of Science as a monument. Of his work as an expositor those who have heard his lectures and attended his class room exercises can speak with unbounded enthusiasm. He was one of the rare men who by presence and voice and manner could bring the truths and ideals of science home even to those pupils with whom scientific study could never be more than an incident in their lives.

But above all his works and above all his qualities stands the figure of Dana himself — more than an explorer, more than a discoverer, more than a teacher; his countenance, as it were, illuminated by a touch of the light of a new day for which the world was being prepared.

> "His life was gentle; and the elements
> So mixed in him that Nature might stand forth
> And say to all the world, 'This was a man.'"

SPENCER FULLERTON BAIRD

Born, Reading, Pa., February 3, 1823
Died, Woods Holl, Mass., August 19, 1887

Zoölogist

Noted for his work in the Smithsonian Institution and the United States Fish
Commission

SPENCER FULLERTON BAIRD.

By Hugh M. Smith.

The life, the character, the work of Spencer Fullerton Baird entitle him to recognition in any assemblage and on any occasion where honor is paid to those who have been their country's benefactors through illustrious achievements in science.

Developing a taste for scientific pursuits at a very early age, and confirmed in those pursuits through the influence of friendships with Agassiz, Audubon, Dana and other leading scientists of the time, Baird was selected as assistant secretary of the Smithsonian Institution when only twenty-seven years old, and there entered on a career devoted to the promotion, diffusion and application of scientific knowledge among men, and marked by dignity, sound judgment, fidelity to duty, versatility and general usefulness.

In the many phases of his intellectual development he resembled Franklin and Cope; in the multiplicity of his public duties and in the diversity of the scientific accomplishments in which he attained eminence he had few equals; in founding, organizing and simultaneously directing a number of great national scientific enterprises he was unique among those whose memory is here extolled today.

To render an adequate account of the branches of scientific endeavor in which he achieved prominence, benefited his own and future generations and added to his country's renown, one would need to be an ornithologist, a mammalogist, an ichthyologist, a herpetologist, an invertebrate zoölogist, an anthropologist, a botanist, a geologist, a palæontologist, a deep-sea explorer, a fishery expert, a fish-culturist, an active administrator of scientific institutions and an adviser of the federal government in scientific affairs,— for Baird was all these and more.

We freely acknowledge today the debt that science owed Baird alive and now owes his memory, especially for his inestimable services as assistant secretary and later as secretary of the Smithsonian Institution, as director of the National Museum and as head of the Commission of Fish and Fisheries. Among all the establishments with which he was connected, this last was preëminently and peculiarly his own.

It was conceived by him and created for him, and it would almost appear that he was created for it, for certainly no other person of his day and generation was so admirably fitted for the task of organizing this bureau and of executing the duties that grew out of its functions as successively enlarged by Congress. Insisting on scientific investigations and knowledge as the essential basis for all current and prospective utilitarian work, he drew around him a corps of eminent biologists and physicists; he established laboratories; he laid plans for the systematic study of our interior and coastal waters; he had vessels built that were especially designed and equipped for exploration of the seas. While he thus inaugurated operations which have been of lasting benefit to the fisheries, at the same time he became the foremost promoter and exponent of marine research; and the knowledge we today possess of oceanic biology and physics is directly or indirectly due to Baird more than to any other person. The rapid development of piscicultural science under his guidance gave to the United States the foremost place among the nations in maintaining and increasing the aquatic food supply by artificial means; and it was no perfunctory tribute when in 1880, at the International Fishery Exhibition held in Berlin, Emperor William awarded the grand prize to Baird as "the first fish-culturist in the world."

The spirit of Baird influences the Bureau of Fisheries today, as it does all the other institutions with which he was associated; and since his death nearly twenty years ago, the good that has been accomplished in the interest of fish culture and the fishing industry, and in the conduct and encouragement of scientific work, has been in consequence of the foundations he laid, the policy he enunciated and the example he set.

But conspicuous as were his services to science and mankind; faithful and unselfish as was his devotion to the executive responsibilities imposed on him; beautiful as was his personal character, I conceive that his most enduring fame may result from the enthusiasm with which he inspired others, and the encouragement and opportunity that he afforded to all earnest workers. The recipients of his aid can be numbered by hundreds, and many of them are today his worthy successors in various fields; and their places in turn will gradually be taken by a vast number of men and women who will perpetuate his memory by efficiently and reverently continuing his work.

This evidence of the donor's beneficence is a noble and impressive

JOSEPH LEIDY

Born, Philadelphia, September 9, 1823
. Died, Philadelphia, April 30, 1891

Anatomist, zoölogist, palæontologist

Noted for pioneer work among the fossil vertebrates of western United States

memorial of one who merited his country's profoundest gratitude; but the bust signifies something more, for it is a recognition of that zeal, fidelity, self-sacrifice, intelligence and strength in the American character so preëminently typified by Spencer Fullerton Baird.

JOSEPH LEIDY.

BY WILLIAM KEITH BROOKS.

Joseph Leidy was born in Philadelphia; there he passed his three score years and ten, and there he died. For forty-five years he was an officer in the Philadelphia Academy of Natural Science, and for forty years a professor in the University of Pennsylvania. His character was simple and earnest, and he had such a modest opinion of his talents and of his work, that the honors and rewards that began to come to him in his younger days, from learned societies in all parts of the world, and continued to come for the rest of his life, were an unfailing surprise to him.

His knowledge of anatomy and zoölogy and botany and mineralogy was extensive and accurate and at his ready command. Farmers and horticulturists came to him and learned how to check the ravages of destructive insects; physicians sent rare or new human parasites and were told their nature and habits and the best means of prevention; jewelers brought rare gems and learned their value. His comments, at the Academy, on the recent additions to its collections gave a most impressive illustration of his ready command of his vast store of knowledge of natural history.

Leidy wrote no books, in the popular meaning of the word. He undertook the solution of no fundamental problem of biology. There are few among his six hundred publications that would attract unscientific readers, or afford a paragraph for a newspaper. They are simple and lucid and to the point. Most of them are short, although he wrote several more exhaustive monographs. They cover a wide field, but most of them fall into a few groups. Many deal with the parasites of mammals — among them, one in which his discovery of trichena in pork is recorded.

Two hundred and sixteen, or about one third, of his publications are

on the extinct vertebrates of North America. His first paper on palæontology was published in 1846, and his last in 1888, as the subject
occupied him for more than forty years. He laid, with the hand of a
master, the foundation for the palæontology of the reptiles and mammals
of North America, and we know what a wonderful and instructive and
world-renowned superstructure his successors have reared upon his
foundation. It was this work that established his fame and brought
his honors and rewards. They who hold it to be his best title to be
enrolled among the pioneers of science in America are in the right, in so
far as the founder of a great department of knowledge is most deserving
of commemoration; but I do not believe it was his most characteristic
work.

I can mention but one of the results of his study of American fossils.
He showed, in 1846, that this continent was the ancestral home of the
horse, and he sketched, soon after, the outline of the story of its evolution
which later workers have made so familiar.

More than half his papers are on a subject which seems to me to
contain the lesson of his life. Like Gilbert White, he was a home
naturalist, devoted to the study of the natural objects that he found
within walking distance of his home, but he penetrated far deeper into
the secrets of the living world about him than White did, finding new
wonders in the simplest living being. In the intestine of the cock-roach,
and in that of the white ant, he found wonderful forests of microscopic
plants that were new to science, inhabited by minute animals of many
new and strange forms. His beautifully illustrated memoir on *A
Flora and Fauna Within Living Animals* is one of the most remarkable
works in the whole field of biological literature. Another memoir
gives the results of his study of the anatomy of snails and slugs. The
inhabitants of the streams and ponds in the vicinity of his home furnished an unfailing supply of material for research and discovery, and
many of his publications are on aquatic animals. He finally became
so much interested in the fresh-water rhizopods that he abandoned all
other scientific work in order to devote his attention exclusively to these
animals. His results were published in the memoir on *The Fresh-
water Rhizopods of North America*. This is the most widely known
of his works. It is, and must long be, the standard and the classic
upon its subject. I have no time to dwell upon his work as the naturalist
of the home — his best and most characteristic work. Its lesson to

EDWARD DRINKER COPE

Born, Philadelphia, July 28, 1840
Died, Philadelphia, April 12, 1897

Palæontologist, biologist, philosopher

Noted for   discoveries among the vertebrate fossils of western United States and
his deductions from their study

later generations of naturalists seems to me to be that one may be useful to his fellow-men and enjoy the keen pleasure of discovery and come to honor and distinction, without visiting strange countries in search of rarities, without biological stations and marine laboratories, without the latest technical methods, without grants of money, and, above all, without undertaking to solve the riddles of the universe or resolving biology into physics and chemistry.

If one have the simple responsive mind of a child or of Leidy, he may, like Leidy, "find tongues in trees, books in the running brooks, sermons in stones and good in everything."

EDWARD DRINKER COPE.

By Henry Fairfield Osborn.

In the marble portrait of Edward Drinker Cope, you see the man of large brain, of keen eye and of strong resolve, the ideal combination for a life of science, the man who scorns obstacles, who while battling· with the present looks above and beyond. The portrait stands in its niche as a tribute to a great leader and founder of American palæontol-- ogy, as an inspiration to young Americans. In unison with the other portraits its forcible words are: "Go thou and do likewise."

Cope, a Philadelphian, born July 28, 1840, passed away at the early age of fifty-seven. Favored by heredity, through distinguished ancestry of Pennsylvania Quakers, who bequeathed intellectual keenness and a. constructive spirit. As a boy of eight entering a life of travel and observation, and with rare precocity giving promise of the finest qualities· of his manhood. Of incessant activity of mind and body, tireless as an explorer, early discovering for himself that the greatest pleasure and stimulus of life is to penetrate the unknown in Nature. In personal character fearless, independent, venturesome, militant, far less of a Quaker in disposition than his Teutonic fellow citizen Leidy. Of enormous productiveness, as an editor conducting the *American Natural-* *ist* for nineteen years, as a writer leaving a shelf-full of twenty octavo and three great quarto volumes of original research. A man of fortitude, bearing material reverses with good cheer, because he lived in the·

world of ideas and to the very last moment of his life drew constant refreshment from the mysterious regions of the unexplored.

In every one of the five great lines of research into which he ventured, he reached the mountain peaks where exploration and discovery guided by imagination and happy inspiration gave his work a leadership. His studies among fishes alone would give him a chief rank among zoölogists, on amphibians and reptiles there never has been a naturalist who has published so many papers, while from 1868 until 1897, the year of his death, he was a tireless student and explorer of the mammals. Among animals of all these classes his generalizations marked new epochs. While far from infallible, his ideas acted as fertilizers on the minds of other men. As a palæontologist, enjoying with Leidy and Marsh that Arcadian period when all the wonders of our great West were new, from his elevation of knowledge which enabled him to survey the whole field with keen eye he swooped down like an eagle upon the most important point.

In breadth, depth and range we see in Cope the very antithesis of the modern specialist, the last exponent of the race of the Buffon, Cuvier, Owen and Huxley type. Of ability, memory and courage sufficient to grasp the whole field of natural history, as comparative anatomist he ranks with Cuvier and Owen; as palæontologist with Owen, Marsh and Leidy — the other two founders of American palæontology; as natural philosopher less logical but more constructive than Huxley. America will produce men of as great, perhaps greater genius, but Cope represents a type which is now extinct and never will be seen again.

Guide Leaflet No. 25

AMERICAN MUSEUM OF NATURAL HISTORY

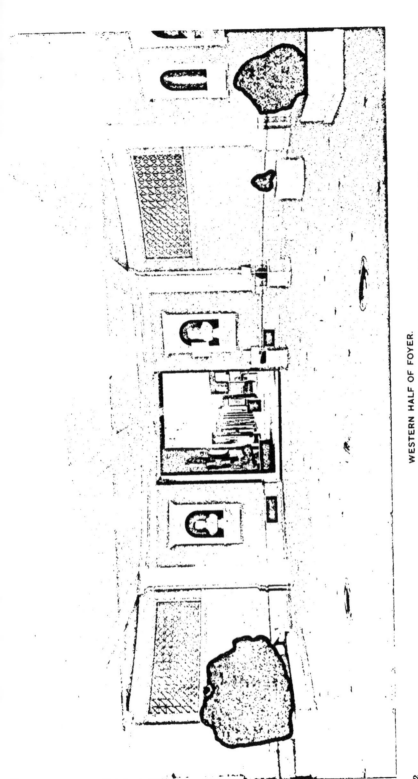

WESTERN HALF OF FOYER.

Showing the following Meteorites : Ahnighito, Brenham, Forest City, The Dog, The Woman.

THE METEORITES

IN THE FOYER OF THE

American Museum

OF

NATURAL HISTORY

By EDMUND OTIS HOVEY, Ph. D.

ASSOCIATE CURATOR OF GEOLOGY

NO. 26

OF THE

GUIDE LEAFLET SERIES

OF THE

AMERICAN MUSEUM OF NATURAL HISTORY

EDMUND OTIS HOVEY, EDITOR

New York. Published by the Museum. December, 1907

NORTH

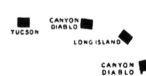

THE FOYER. NO. 104.

First (Ground) Floor, Central Section of Building.

TABLE OF CONTENTS.

LIST OF ILLUSTRATIONS.

THE METEORITES IN THE FOYER OF THE AMERICAN MUSEUM OF NATURAL HISTORY.[1]

By Edmund Otis Hovey, Ph. D.

Associate Curator of Geology.

Introduction.

SCARCELY a century ago the scientific and even the popular world scoffed at the idea that masses of matter could possibly come from outer space (or "heaven") and strike the surface of the earth,— in other words that stones could fall from the sky. Even at the present time, although it is well known that occasionally masses of metal and stone — "meteorites" — do fall from the sky, there is much misinformation current in regard to their character and the conditions under which they have come to the earth.

Livy, Plutarch and other early historians mention several stones which had been seen to fall from the sky. Among these were a stone which fell in Phrygia and was kept there for centuries until it was removed to Rome about 204 B. C. with imposing ceremonies; a shower of stones that fell in the Alban Mountains near Rome about 652 B. C., and a stone that fell in Thrace in the fifth century B. C. and was known to Pliny five hundred years later. The image of the goddess Diana which was preserved at Ephesus is said to have "fallen down from Jupiter" and was probably a meteorite, and idol known as the Venus of Cyprus seems likewise to have had the same origin. Stones which have fallen from the sky have been regarded as being of miraculous origin and have been worshiped by many primitive peoples. They have been viewed with awe too by tribes and nations which could not be considered primitive, including some in India, China and Japan.

Arguments which form strange reading at the present day were advanced by eminent scientists against the idea that any bodies could come to the earth from space, and French scientists were particularly

[1] *Guide Leaflet No. 26* of the American Museum series.

vehement in their denial of such origin. Even the famous chemist Lavoisier was one of a committee of three who presented to the French Academy in 1772 a report upon a stone, the fall of which was said to have occurred at Lucé four years previously. They recorded their opinion that the stone was an ordinary one which had been struck by lightning. It was, nevertheless, a true meteorite.

Early in the year 1794 Professor Chladni, a renowned German physicist, published a thesis in which he collated many accounts of bodies which had been said to have fallen from the sky, discussed the nature of the bodies themselves and expressed the conviction that bodies could and did come to our earth from space. Chladni devoted particular attention to the iron-and-stone mass known as the "Medwedewa" meteorite and the iron mass known as Campo del Cielo. The former of these was first described by the traveler Pallas, who saw it in the year 1772 at the city of Krasnojarsk, Siberia. The latter was found by Indians in the interior of Argentina, South America, and was first visited in the year 1783 by Don Michael Rubin de Celis, who calculated the weight of the mass to be 30,000 pounds.

As if in direct confirmation of Chladni's theory, a shower of stones fell at Siena, Italy, on June 16, 1794, and the occurrence is thus described in connection with the account of an eruption of Vesuvius by Sir William Hamilton[1]:

"I must here mention a very extraordinary circumstance indeed, that happened near Sienna in the Tuscan state, about 18 hours after the commencement of the late eruption at Vesuvius on the 15th of June, though that phenomenon may have no relation to the eruption; and which was communicated to me in the following words by the Earl of Bristol, bishop of Derry, in a letter dated from Sienna, July 12th, 1794: 'In the midst of a most violent thunder-storm, about a dozen stones of various weights and dimensions fell at the feet of different people, men, women, and children; the stones are of a quality not found in any part of the Siennese territory; they fell about 18 hours after the enormous eruption of Vesuvius, which circumstance leaves a choice of difficulties in the solution of this extraordinary phenomenon: either these stones have been generated in this igneous mass of clouds, which produced such unusual thunder, or, which is equally incredible, they were thrown from Vesuvius at a distance of at least 250

[1] Philosophical Transactions of the Royal Society of London. Abr. ed., 1809, vol. XVII, p. 503.

miles; judge then of its parabola. The philosophers here incline to the first solution. I wish much, Sir, to know your sentiments. My first objection was to the fact itself; but of this there are so many eye-witnesses, it seems impossible to withstand their evidence, and now I am reduced to a perfect scepticism.' His lordship was pleased to send me a piece of one of the largest stones, which when entire weighed upwards of 5 lb.; I have seen another that has been sent to Naples entire, and weighs about 1 lb. The outside of every stone that has been found, and has been ascertained to have fallen from the cloud near Sienna, is evidently freshly vitrified, and is black, having every sign of having passed through an extreme heat; when broken, the inside is of a light grey color mixed with black spots, and some shining particles, which the learned here have decided to be pyrites, and therefore it cannot be a lava, or they would have been decomposed."

Scientists, however, are often hard to convince, and some of that day contended that the Siena stone had been formed in the air by condensation of the particles of dust in an eruption cloud from Vesuvius, in spite of the fact that Siena is 250 miles distant from the volcano and that the largest stone of the shower weighed $7\frac{1}{2}$ pounds, while several weighed more than 1 pound each. Even the 56-pound stone which fell December 13, 1795, at Wold Cottage near Scarborough, Yorkshire, England, almost at the feet of a laborer, did not dislodge this theory from the mind of Edward King, its originator.

The cloud theory was completely disproved at Krakhut near Benares, India, on December 19, 1798, when, directly after the passage of a ball of fire through the air, a heavy explosion or a series of explosions was heard and many stones[1] fell from a sky which had been perfectly cloudless for a week before the event and remained so for many days afterward. Even these facts, however, did not fully convince the scientists of France and it required the occurrence of the meteoritic shower of L'Aigle, France, April 26, 1803, for final proof.[2] L'Aigle is easily accessible from Paris and M. Biot, a noted physicist was sent at once to investigate the matter. Biot learned that on the day mentioned a violent explosion occurred in a practically clear sky in the vicinity of

---

[1] Represented in the general meteorite collection, Morgan Hall, No. 404 of the fourth floor of this building, by a small fragment one fourth of an ounce (8 grammes) in weight.

[2] A fragment of L'Aigle weighing 5 ounces (157 grammes) is in the general collection.

L'Aigle which was heard over an area seventy five miles in diameter directly after a swiftly moving fire-ball had been seen to pass through the air. The explosion, or series of explosions, was immediately followed by the fall of two or three thousand stones within an elliptical area about 6¼ miles long and 2½ miles wide. The largest of the stones weighed 20 pounds, the next largest 3½ pounds, but most of the fragments were very small. The occurrence at L'Aigle proved the correctness of another of Chladni's theories, which was that " fire balls " in the sky were nothing more or less than meteorites in flight.

The oldest still existing meteorite of the fall of which we have an exact record is that of Ensisheim, in Elsass, Germany.[1] An ancient document states:

"On the sixteenth of November, 1492, a singular miracle happened: for between 11 and 12 in the forenoon, with a loud crash of thunder and a prolonged noise heard afar off, there fell in the town of Ensisheim a stone weighing 260 pounds. It was seen by a child to strike the ground in a field near the canton called Gisguad, where it made a hole more than five feet deep. It was taken to the church as being a miraculous object. The noise was heard so distinctly at Lucerne, Villing, and many other places, that in each of them it was thought that some houses had fallen. King Maximilian, who was then at Ensisheim, had the stone carried to the castle; after breaking off two pieces, one for the Duke Sigismund of Austria and the other for himself, he forbade further damage, and ordered that the stone be suspended in the parish church." [2]

Within the past century many stones and some masses of iron have been seen to fall from the sky and afterwards have been collected and are now in cabinets, while several hundred specimens have been found which are so much like the positively known meteorites that they have been classed with them and are jealously guarded in collections.

Classification.

Meteorites are generally divided into three classes according to their mineral composition:

1. "Siderites," or iron meteorites, which consist essentially of an alloy of iron and nickel;

[1] A fragment of this meteorite weighing about four ounces (129 grammes) is in the general meteorite collection.
[2] Fletcher. An Introduction to the study of Meteorites. P. 19. 1888.

AHNIGHITO, THE GREAT CAPE YORK METEORITE.

Weight, more than 36.5 tons. The largest and heaviest meteorite known.

2. "Siderolites," or iron-stone meteorites, which are formed of a nickel-iron sponge, or mesh, containing stony matter in the interstices;

3. "Aërolites," or stone meteorites, which are made up mainly of stony matter, but almost always contain grains of nickel-iron scattered through their mass.

The line of demarcation between these classes is not always sharp, and there are many subordinate kinds of aërolites.

Countless numbers of meteoritic bodies, mostly of minute size, must exist within the boundaries of the solar system, since from fifteen to twenty millions of them enter the earth's atmosphere every day. Almost all of these are dissipated in our atmosphere through heat produced by friction with the air, so that the only evidence of their presence is a trail of light across the sky. This usually is visible only at night, and is familiar to all as a shooting star or meteor. Shooting stars are to be seen almost every evening, but they are particularly abundant during August and November. Sometimes the November shower of meteors has been so pronounced that the sky has seemed fairly to radiate lines of fire, an effect far surpassing in brilliance the most ambitious artificial fire works. Not one in a hundred million of these shooting stars, however, reaches the earth in a recognizable mass; in fact, there are records of only about 685 known meteorites which are represented in museums and private cabinets.

The weight of known meteorites varies between wide limits. The lightest independent mass is a stone meteorite weighing about one sixth of an ounce called Mühlau from the town in the vicinity of Innsbruck, Austria, near which it was found in 1877; the heaviest mass known is Ahnighito, of the Foyer collection, an iron meteorite weighing more than thirty-six and one half tons which came from Cape York, Greenland. Some showers of meteorites have furnished even smaller individuals than Mühlau. Forest City, well represented in the Foyer collection, has been found in fragments weighing one twentieth of an ounce. Pultusk is a famous fall and the smallest of the "Pultusk peas," as the material is called, weigh less than one thirtieth of an ounce each, while Hessle fell in a veritable rain of meteoritic dust, the smallest particles of which weigh about one four hundred twenty-fifth of an ounce and could never have been found had they not fallen on an ice-covered lake, where they were readily seen and recognized.

Meteoritic masses are almost certainly extremely cold during their existence in outer space, but when they come into the earth's atmosphere friction with the air raises the temperature of the surface to the melting point, producing a great amount of dazzling light as well as superficial heat. In spite of this surface fusion, it is highly probable that the duration of the aërial flight of a meteorite is so short that in many cases the interior does not become even warm.[1]

The rapid heating of the exterior and the differences of temperature between different parts of a meteorite often lead to its rupture before it reaches the ground. This is particularly the case with stone meteorites, the iron meteorites being tough enough usually to withstand the fracturing agencies. Most of the meteorites which have burst have furnished only two or three fragments, so far as known, but a few have furnished many, while there have been found 700 pieces of Hessle, 1000 pieces of Forest City, several thousand each of Knyahinya and L'Aigle, and about one hundred thousand each of Mocs and Pultusk. The name "stone shower" has been appropriately given to the falls comprising many individuals. "Iron showers" from bursting siderites are much rarer than the stone showers, only six are known to have occurred, among which Canyon Diablo leads, several thousands of fragments of this famous fall having been found.

The breaking up of a meteorite is accompanied by an explosion or series of explosions, and often these are startling in their sharpness and intensity, when they occur near the earth. Forest City, 268 pounds of which have been found, just before falling burst in a series of explosions which were heard over an area two hundred miles in diameter. There were three distinct detonations connected with the fall near Butsura, Bengal, which were heard at Goruckpur sixty miles away, although the meteorite was a small one, less than fifty pounds of it having been found. The occurrences at Krakhut, L'Aigle and Ensisheim have already been mentioned.

Chemical Composition.

Some forty one elements, four of which are gases, are said to occur in meteorites, but several of these may be regarded as doubtful. The

---

[1] See also page 18.

most abundant have been arranged by Dr. O. C. Farrington[1] in the following order of importance:

1. Iron   3. Silicon   5. Nickle   7. Calcium
2. Oxygen   4. Magnesium   6. Sulphur   8. Aluminum

The other elements of particular importance in this connection are carbon, chlorine, chromium, cobalt, copper, hydrogen, manganese, nitrogen, phosphorus, potassium and sodium.

Mineral Constituents.

Seven elements have been found in meteorites in the elemental or uncombined state. They are iron, nickel, cobalt and copper in the form of alloys, carbon, hydrogen and nitrogen. With these exceptions, the constituents of meteorites are chemical compounds and all but six of the whole list have their exact equivalents in minerals which are found in the crust of the earth.

According to most authorities the constituents of meteorites may be divided into essential and accessory components as follows[2]:

Essential.	Accessory.
Nickel-iron	*Schreibersite
Olivine (chrysolite)	Diamond
Pyroxenes (Orthorhombic)	Graphite (Cliftonite)
Pyroxenes (Monoclinic)	Hydro-carbons.
Feldspar (Plagioclase)	Cohenite
*Maskelynite.	*Moissanite
	*Troilite
	Pyrrhotite
	*Daubréelite
	*Oldhamite
	Tridymite
	Chromite
	Magnetite
	Osbornite
	Lawrencite
	Glass

---

[1] Journal of Geology. Vol. IX, p. 394. 1901.

[2] E. Cohen. Meteoritenkunde I, p. 322. 1894. The asterisk indicates the minerals which are peculiar to meteorites and are not known to occur in the earth's crust.

Essential Constituents.

The iron of meteorites is always alloyed with from 6 to 20 per cent of nickel. This " nickel-iron," as it is commonly called, is usually crystalline in texture, and when it is cut, polished and "etched" a beautiful network of lines is brought out, indicating plates which lie in positions determined by the crystalline character of the mass. This network of lines constitutes what are called the Widmanstätten figures, from the name of their discoverer. When these figures are strongly developed, the meteoritic origin of the iron cannot be questioned, but their absence does not necessarily disprove such an origin. Native iron of terrestrial

WIDMANSTÄTTEN LINES, OR FIGURES

Carleton Iron Meteorite. Natural size. In this iron the plates are very thin.

origin is extremely rare and has been found almost exclusively at Disco Island and immediate vicinity on the west coast of Greenland. The Disco, or Ovifak, iron contains less nickel than meteoritic iron, while other terrestrial nickel-irons (*i. e.* awaruite and josephinite) contain much more. Small quantities of metallic cobalt are also alloyed with the nickel and a little copper is sometimes found in the same association.

Next to nickel-iron the mineral olivine, or chrysolite, is the most important constituent. This is a silicate of magnesium, always con-

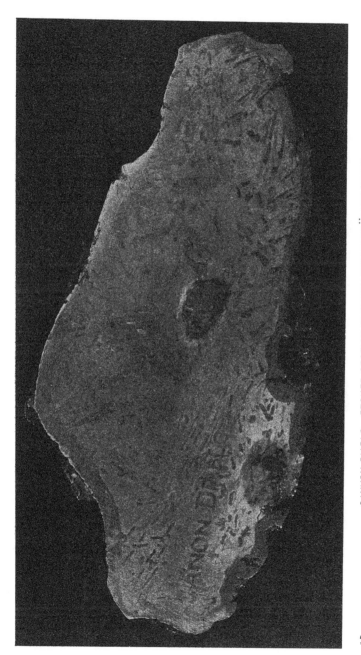

CANYON DIABLO. ETCHED SECTION SHOWING WIDMANSTÄTTEN LINES.

A diamond was found in the black spot in this section.

15

taining some iron, which occurs in all the siderolites and most of the aërolites, sometimes comprising a considerable portion of their mass. It is a dark yellowish-green to black, glassy mineral usually occurring as rounded or angular grains, but sometimes as crystals. It is prominent in a slice of Brenham in the Foyer collection, where it forms glassy grains in a mesh of nickel-iron. Olivine is the gem, peridot.

The minerals belonging to the group known as orthorhombic pyroxenes are next to olivine in point of abundance. Chemical analyses show that all gradations are present from the colorless enstatite to the almost black hypersthene members of the group. The monoclinic pyroxenes, which are important constituents of terrestrial igneous rocks, are represented in meteorites by only two forms, an iron-alumina pyroxene like common augite and one nearly free from iron and without alumina which is to be compared with diopside. The augite-like mineral is brown to green in color and occurs usually in grains or splinters rarely in crystals. It has been found in many meteorites, but diopside has been identified only once with certainty.

The great feldspar series has been identified in meteorites in four of its forms, namely: anorthite, albite, oligoclase and labradorite. Of these, anorthite has been found forming a large part (35 per cent.) of some meteorites and measurable crystals have been obtained, but in most cases where feldspar occurs in a meteorite, it has been possible to go no farther than to identify it as belonging to the plagioclase section of the mineral group.

Maskelynite is a transparent, colorless, glassy mineral. In chemical composition it is related to the terrestrial species leucite, but it is a distinct form and thus far is known only from meteorites. It is not known to occur in any of the meteorites displayed in the Foyer.

Accessory Constituents.

Schreibersite is a phosphide of iron, nickel and cobalt which is probably peculiar to meteorites. It is tin-white in color, changing to bronze-yellow or steel gray on exposure to the air. In structure it is granular, flaky, crystalline or needle-like. Next to nickel-iron schreibersite is the most generally disseminated constituent of siderites and forms some of the shining lines to be seen on etched sections.

Carbon occurs in at least three forms in meteorites, as the diamond,

.as graphite (cliftonite) and as hydrocarbons. Diamonds were first
found in Canyon Diablo in 1891. They are extremely minute in size
but recognizable crystals have been obtained. Graphite (cliftonite)
occurs usually in nodules and only in siderites in particles that are large
enough for easy examination. The material is very fine. The cliftonite
form of graphite is considered by most authorities to be a pseudomorph
after diamond.

Hydrocarbons of several kinds have been found in meteorites. According to Cohen[1] they may be grouped into three classes: (1) compounds
of carbon and hydrogen alone; (2) compounds of carbon, hydrogen and
oxygen; (3) compounds of carbon, hydrogen and sulphur. None of
the meteorites in the Foyer collection is known to contain any hydrocarbon, but the fact that any meteorite should contain such substances
is of great scientific interest. It is pretty clear that they belong to the
pre-terrestrial history of the masses; hence, since they are readily com-
bustible or volatile, the meteorites that contain them cannot have been
heated to high temperatures, at any rate, subsequent to the formation
of the compounds. This is an additional argument in support of the
statement already made that the heating of meteorites during aërial
flight is, in many instances at least, only superficial. Furthermore, the
existence of hydrocarbon compounds in meteorites, where no life can
have existed, shows that organisms are not absolutely necessary to the
formation of such compounds in the earth's crust.

Cohenite, which is a carbide of iron, nickel and cobalt, is tin-white
in color and looks like schreibersite. It is much rarer, however, and
occurs in isolated crystals. The only terrestrial occurrence of cohenite
is in the basaltic iron of Greenland. Moissanite, the natural carbide
of aluminum corresponding to the artificial carborundum, has thus far
been found only in Canyon Diablo, where it occurs in microscopic
crystals. It is the latest discovery among the constituents of meteorites,
having been found in 1905 by Henri Moissan.

As far as investigations have been carried, heating develops the fact
that meteorites contain gases condensed within them, either by occlusion
in the same way that platinum and zinc absorb hydrogen or by some
form of chemical union. According to Cohen[2] the following gases have

[1] Meteoritenkunde. Heft I, p. 159.
[2] Meteoritenkunde. Heft 1, p. 169

been found: hydrogen, carbon dioxide, carbon monoxide, nitrogen and marsh gas (light carburetted hydrogen).

Troilite is common in meteorites and constitutes brass- or bronze-yellow nodules, plates and rods which are to be seen in nearly every section, particularly of siderites. The mineral is usually considered to be the simple sulphide of iron, FeS, but its exact chemical composition and crystalline structure are still matters of investigation and dispute. Canyon Diablo and Willamette contain, or contained, much troilite in the shape of rods, and the fusion and dissipation of this mineral during the aërial flight of the masses gave rise to some of the holes which penetrate them, and the same statement is true of many other meteorites. Pyrrhotite, the magnetic sulphide of iron, $Fe_{11} S_{12}$, occurs in stone meteorites and chiefly in the form of grains. Daubréelite is likewise a sulphide of iron, but it differs from those just mentioned through containing much chromium, giving the chemical formula FeS, $Cr_2 S_3$. The mineral is peculiar to siderites and siderolites and has never been found in the stone meteorites or in the earth's crust. It occurs in Canyon Diablo, where it may be seen surrounding nodules of troilite as a black shell with metallic luster.

Oldhamite, a sulphide of calcium, CaS; tridymite, a form of silica, SiO_2; chromite, an oxide of iron and chromium, $FeCr_2 O_4$; magnetite, the magnetic oxide of iron, $Fe_3 O_4$; osbornite, another sulphide or oxysulphide of calcium, and lawrencite a chloride of iron, $Fe Cl_2$, occur sparingly in some meteorites. Lawrencite manifests itself rather disagreeably through alteration to the ferric chloride, which oozes out of the masses of iron and stands in acrid yellow drops on the surface or runs in streaks to the bottom. Glass like the volcanic glass of terrestrial rocks seems never to be absent from the interior of stone meteorites, but from the nature of the case it is not found in iron meteorites.

Surface Characteristics.

The surface of a newly-fallen meteorite always consists of a thin veneer or crust which differs in marked degree from the interior of the mass. In the case of the siderites, this seems to be a polish due to melting and friction, together with partial oxidation. Some iron meteorites which are known to have lain long in the ground likewise show a crust which is somewhat similar in appearance, but it is due to slow oxidation or "rusting" in the ground and is called the rust crust. Almost without exception the aërolites are covered with a crust, the appearance of which varies according to the mineral composition of the mass. The crust is almost always black and is usually dull, but sometimes it has high luster. A few meteorites possess a dark-gray crust, and some show crust only in patches.

The crust of the stone meteorites is glassy in character on account of its being composed of silicates which have been cooled rapidly from fusion. This glass, like glasses of volcanic origin, does not long resist the atmospheric agents of decay, hence it is usually missing from those aërolites which have lain long in the earth or it is much decomposed, as may be seen by examining Long Island and Selma in the Foyer collection. The crust varies in thickness on different parts of a meteorite and often shows ridges and furrows which are due to friction with the air. Frequently the ridges or furrows radiate from one or more centers in such manner as to show which side of the mass was forward during its flight through the air. So quickly is the crust formed that even the smallest members of a meteorite shower usually possess a complete crust. In the case of angular fragments the crust on the different sides can usually be distinguished as "primary" or "secondary" according to whether it was a part of the original exterior of the mass or was formed upon the new surfaces exposed by the bursting of the meteorite.

Another common surface characteristic of meteorites is an abundance of shallow depressions or pittings, which on account of their form have been called "thumb marks," or piëzoglyphs. These pittings are so shallow and superficial in character that exposure to the atmosphere obscures or obliterates them in a comparatively short time. The rusting of an iron meteorite may produce similar shallow depressions, as will be seen from an examination of the surfaces of the great hollows in Willamette. The true piëzoglyphs doubtless owe their origin to several different causes, the most potent of which are unequal softening of a mass due to varying chemical composition and rapidly changing pressure and consequent erosion during flight through the atmosphere.

Without going more deeply into the subject in general we may now turn our attention to the Foyer collection.

CAPE YORK.
"The Woman". Weighs 3 tons.

THE CAPE YORK METEORITES,

"AHNIGHITO," OR "THE TENT," "THE WOMAN" AND "THE DOG."

(Siderites.)

For centuries, and perhaps for thousands of years, the three masses of iron known as the Cape York meteorites lay on the north coast of Melville Bay near Cape York, Greenland, but they were seen for the first time by a white person, when they were visited by Commander Robert E. Peary, U. S. N., in 1894 and 1895 under the guidance of Tallakoteah, a member of the Eskimo tribe which up to the early part

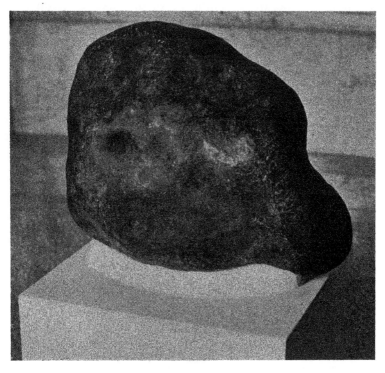

CAPE YORK.
"The Dog". Weighs 1100 pounds.

of the nineteenth century had obtained material for knives and other utensils from the masses.

The three meteorites were known as a group to the Eskimo under the name of "Saviksue" or "The Great Irons," and each had its own name suggested by its shape. The smallest mass, weighing about 1,100 pounds, was called "The Dog"; the next larger mass, weighing about three tons, was named "The Woman," because the shape was thought to suggest the squatting figure of a woman with a babe in her arms and a shawl thrown about her, and the largest mass, weighing more than thirty six and one half tons, was known as "The Tent." The last, however, has been formally christened by the daughter of the explorer with her own name, "Ahnighito." This great mass is 10 feet 11 inches long, 6 feet 9 inches high and 5 feet 2 inches thick.

The Woman and the Dog were visited by Peary in 1894 and were obtained the following year after much difficult and exciting work, an incident of which was the breaking up of the cake of ice on which the Woman had been ferried from the shore to ship just as the mass was about to be hoisted aboard. Fortunately there was enough tackle around the meteorite to prevent its loss. In 1895, Commander Peary visited Ahnighito, also, which lay on an island only four miles from the two smaller masses, but he could do little toward its removal. The next year he made another voyage for the purpose of getting the great iron but was unsuccessful. His third attempt was made in 1897, and the meteorite was brought safely to New York in the ship "Hope."

SECTION OF AHNIGHITO NATURAL SIZE.
Shows broad Widmanstätten lines.

The three masses are closely similar in chemical composition, analyses by J. E. Whitfield giving the following results:

	The Dog.	The Woman.	Ahnighito.
Iron	90.99%	91.47%	91.48%
Nickel	8.27%	7.78%	7.79%
Cobalt	0.53%	0.53%	0.53%

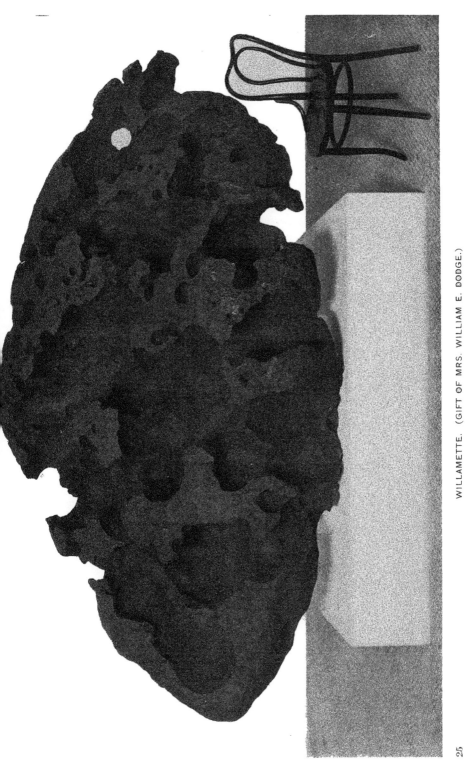

WILLAMETTE. (GIFT OF MRS. WILLIAM E. DODGE.)

Length 10 feet, height 6 feet 6 inches. Weight 15.6 tons. Rear side, showing deep pits formed by oxidation.

25

Besides these metals there are present small quantities of copper, sulphur, phosphorus and carbon. The similarity in chemical composition and the close proximity in which the masses lay when found indicate the probability that they are parts of the same fall.

WILLAMETTE

(Siderite.)

GIFT OF MRS. WILLIAM E. DODGE.

This is the most interesting iron meteorite, as to external characteristics, which has ever been discovered, and it is the largest ever found in the United States. Its chief dimensions are, length 10 ft., height

SECTION OF WILLAMETTE.

6 ft. 6 in., thickness, 4 ft. 3 in. On the railroad scales in Portland, Oregon, the net weight was shown to be 31,107 lbs.

Willamette was discovered in the autumn of 1902 in the forest about nineteen miles south of Portland, by two prospectors who were searching for ledges likely to contain mineral wealth, particularly gold or silver.

The finders at first supposed that they had come upon a ledge of solid iron, but the meteoritic character of the mass was soon ascertained. Later one of the prospectors removed the meteorite to his own ranch three-quarters of a mile distant, but the owners of the land on which it had been found instituted suit for its recovery, and the contest was carried to the supreme court of the State before the finder relinquished his claim. The specimen was received at the Museum in April, 1906.

The most striking characteristic of Willamette, next to its size, is the series of hollows and deep pits which indent its surface. The broad shallow hollows on the front side, "brustseite," (side now turned toward the wall) were probably caused by friction against the atmosphere and consequent melting and flowage of the iron during the flight of the meteorite through the air. The deep pot-like pits on the rear side (the side now facing the center of the Foyer) are most probably due to rusting while the meteorite was lying in the ground where it fell, and they seem to have had their origin in the decomposition of spheroidal nodules of troilite. Note also the cylindrical holes which penetrate deeply into the mass from both sides. These probably began with the decomposition of rod-like masses of troilite. In addition to these holes and pits the surface of the mass is indented with small shallow depressions which also seem to be a feature of the decomposition of the iron.

A fractured face shows Willamette to be remarkable for its coarse granular texture, the grains being bounded by almost definite planes suggesting crystals. A polished and etched surface shows rather broad Widmanstätten lines. Chemical analysis shows that the meteorite contains about 91.55 per cent iron, 8.09 per cent nickel and a small amount of cobalt, phosphorus and sulphur.

CANYON DIABLO.

(Siderite.)

Canyon Diablo is a siderite which is popularly famous chiefly from the fact that it contains diamonds. This gem stone has been definitely proven to occur in only two meteorites, the other being a Russian fall, although many masses are known to contain carbon in the form of a soft black powder. The discovery of diamonds in Canyon Diablo was made in 1891 by Professor G. A. Kœnig of Philadelphia and was afterward confirmed by Dr. George F. Kunz of New York, Professor Huntington of Harvard University, Professor Moissan of Paris and other investigators. In 1905 Moissan dissolved a fragment of Canyon Diablo weighing

CANYON DIABLO.
Weight, 1087 pounds. Diamonds have been found in specimens of this fall.

several pounds and obtained not only recognizable crystals of the diamond, but also crystals of a mineral corresponding exactly in composition to the extremely hard artificial silicide of carbon (CSi_2) known as carborundum. This new mineral has been named moissanite, and this is the first time that it has been found in nature.

Canyon Diablo was found in 1891 at and near Coon Butte, Arizona, in the vicinity of the town of Canyon Diablo. The original size of the mass is not known, but thousands of fragments have been collected, varying in weight from a fraction of an ounce up to 1,087 pounds. More than 16 tons of this material are said to have been found within the radius of 2½ miles of Coon Butte. Coon Butte is a conical hill rising from 130 to 160 feet above the surrounding plain and containing a crater-like hollow about three-quarters of a mile in greatest diameter and probably 1,460 feet deep originally.

There is no lava of any kind in Coon Butte or in its immediate vicinity, and it is now supposed to be most probable that the "crater" was caused by an immense meteorite striking the earth at this point. The main portion of the mass has not yet been discovered, the fragments which have thus far been found being only the portions separated from the original mass during its passage through the atmosphere and at the time of its impact with the earth.

Two fragments of Canyon Diablo are in the Foyer collection, one of which weighs 1,087 pounds and is the largest piece which has been discovered. It was described and figured by Professor Huntington in the Proceedings of the American Academy of Arts and Sciences, Boston, for 1894. A slice of the meteorite in which a diamond was found is in the general Museum collection and is figured on page 15.

A polished and etched section shows strong Widmannstätten lines which are comparatively broad and somewhat discontinuous. The meteorite consists of 91.26 per cent iron and 8.25 per cent nickel and cobalt, with small quantities of copper, platinum, iridium, phosphorus, sulphur, carbon and silicon. Nodules of troilite are abundant in some parts of the masses. Through decomposition and erosion these nodules have given rise to deep holes in the iron.

TUCSON.[1]

(Siderite.)

The Tucson meteorite, which is also known as the "Signet" or the

---

[1] This specimen is a reproduction in cast iron of the famous Tucson meteorite, the original of which is in the National Museum at Washington. The model from which this reproduction has been prepared was presented to the American Museum by the Smithsonian Institution. The original weighs 1,400 pounds, and this cast has the same weight.

"Ring" on account of its peculiar shape, was found in the Santa Catarina Mountains about thirty miles northwest of the city of Tucson, Arizona, and its existence was known to the Spaniards for at least two hundred years before the region became part of the United States. Tradition, indeed, relates that this and many other fragments fell in a single meteoritie shower about the middle of the seventeenth century. The attention of Americans was first drawn to this and its mate, the Carleton meteorite,

TUCSON, OR "SIGNET."
Weight, 1400 pounds. A cast in iron.

in 1851 by Professor John L. Leconte, who described them as being in use by the blacksmiths of the town as anvils. In 1863 Signet was taken to San Francisco and thence transported by way of the Isthmus of Panama to the Smithsonian Institution at Washington. Carleton, weighing 632 pounds, had been removed to San Francisco the preceding year and was afterwards deposited in the hall of the Pioneers' Society in that city.

Tucson is classed as a siderite, but its average composition shows the presence of from 8 to 10 per cent of stony matter included in the nickel-iron, and the proportions of the mineral constituents vary considerably in different parts of the mass.   The nickel-iron is an alloy consisting of 89.89 per cent iron, 9.58 per cent nickel, 0.49 per cent cobalt and 0.04 per cent. copper, while the stony matter consists of olivine, carrying an unusual quantity of lime and associated with noteworthy quantities of schreibersite and chromite.

BRENHAM.

(Siderolite.)

Brenham is classed as a siderolite, but some of its fragments are entirely of nickel-iron.   The etched section shown in the Foyer illustrates clearly the peculiar texture of the mass.   The metallic portions consist

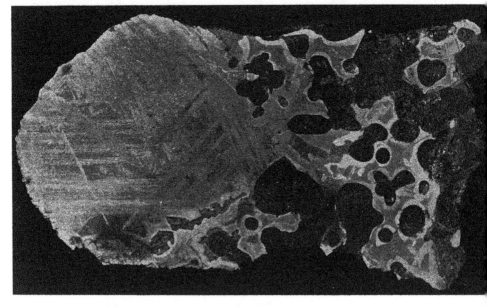

BRENHAM
Siderite (left) and Siderolite (right) in the same piece.

of about 88 per cent. iron, 10 per cent. nickel and 2 per cent. other substances.   The dark green and glassy portions are crystals of olivine, which break out from the iron almost entire.

The Brenham meteorite was found in the year 1886 scattered in many pieces on the prairie in Brenham Township, Kiowa County, Kansas, over an area more than a mile in length. The fragments were hardly covered by the original prairie soil, and several of them were projecting through the sod. Nearly all were found by being struck by mowing machines, plows or other farm implements. The occurrence of heavy "rocks" in a region where stones of any kind are a great rarity was a source of surprise to the ranchmen and led finally to the discovery that they were meteoritic in origin.

About thirty fragments of the meteorite have been found, several of which were used for many purposes about the ranches and had a rather prosaic history before their value was learned. The smaller but heavier (75-pound) mass here exhibited was used for years to hold down a cellar door or the cover of a rain barrel, while the larger but lighter (52.5-pound) mass served as a weight on a hay-stack. It is probable that the meteorite of which these are fragments burst soon after reaching the earth's atmosphere. The total weight of all the fragments of Brenham which have been found is about 2,000 pounds; the largest piece known weighs 466 pounds, the smallest an ounce or two.

Other specimens of this meteorite may be seen in the Morgan Hall of Mineralogy on the Fourth Floor.

FOREST CITY.

(Aërolite.)

On Friday, May 2, 1890, at 5:15 P. M., a brilliant ball of fire shot across the sky from west to east in northern Iowa, its flight being accompanied by a noise likened to that of a heavy cannonading, or of thunder, and by scintillations like those of fireworks. The meteoric light was dazzling even in the full daylight prevailing at the time and the noises, which were due to explosions, were heard throughout a district 200 miles in diameter. This meteor was the Forest City meteorite.

The meteorite burst when it was about 11 miles northeast of Forest City, Winnebago County, whence its name, and most of the fragments were scattered over an area about one mile wide and about two miles long. More than a thousand fragments of this meteorite have been found, most of which weigh from $\frac{1}{20}$ of an ounce to 20 ounces, but a few weigh several pounds. Each is a perfect little meteorite. The largest of the group, which is exhibited here in the Foyer collection, weighs about 75 pounds. The black glassy crust over the surface of all the masses shows that the meteorite exploded early enough in its atmospheric flight for even the smallest fragments to become superficially fused by friction with the air. The fragments show a "primary" and

FOREST CITY.

Shows crust on large and small pieces.

a "secondary" crust, the former formed before and the latter after the bursting of the original mass.

Forest City consists essentially of feldspar, enstatite (a member of the orthorhombic-pyroxene group of minerals), graphite, troilite and nickel-iron. The iron is present in small particles disseminated through the masses and in definite lines suggesting the Widmanstätten figures of a siderite.

The approximate mineral composition of Forest City is

      Nickel-iron......................19.4%

      Troilite...........................6.2%

      Silicates (feldspar, enstatite, etc.)....74.4%

The nickel-iron is an alloy consisting of

           Iron.............92.7%

           Nickel...........6.1%

           Cobalt...........0.7%

The specific gravity of the mass is 3.8. Some chromite is present, but not as much in proportion as is found in the Long Island, Kansas, meteorite.

Some of the smaller individuals of this fall may be seen in the general Museum collection on exhibition in the Morgan Hall of mineralogy (No. 404 of the fourth floor).

LONG ISLAND (KANSAS).

(Aërolite.)

Long Island is the largest stone meteorite known, the fragments which have been recovered aggregating more then 1,325 pounds in weight. The pieces here exhibited weigh together 86 pounds, the largest weighing 32½ pounds. Some of them show the original external surface of the meteorite, but most of them show only fractures. The meteorite was found in more than 3,000 pieces scattered over a gourd-shaped area only 15 or 20 feet long and 6 feet wide in the northwestern corner of Phillips County, Kansas, near the town of Long Island, whence its name. The small area of distribution shows that the mass burst just as it struck the ground, or that it was broken by impact. The late time of bursting is also indicated by the lack of secondary crust on the pieces.

Stony matter makes up about 80 per cent by weight of Long Island, the remainder having originally been nickel-iron and troilite, now partly changed to limonite through rusting. On the polished surfaces of some of the fragments in the case the nickel-iron may be seen as small shining dots. The stony matter consists essentially of the minerals bronzite (one of the orthorhombic pyroxenes), olivine and chromite and bears a close and interesting resemblance to the terrestrial basaltic-igneous rock peridotite. The content of chromite (9 per cent of the whole) is remarkable and is the highest yet reported in meteorites.

Long Island presents a feature heretofore unknown in meteorites. Certain of the planes of fracture show striated surfaces with grooving and polishing (slickensides) due to the parts grinding together in their

LONG ISLAND

Slickensided surface showing movement in the mass before it fell.

flight through space before the mass reached our atmosphere. Two of the pieces in this case show such slickensided surfaces and one of them is illustrated on this page.

Other fine specimens of Long Island may be seen in the general meteorite collection on the fourth floor of the Museum (Hall No. 404).

SELMA.

Weight, 306 pounds.   Front or " Brustseite."

SELMA.

(Aërolite.)

The Selma meteorite is believed to have fallen at about 9 o'clock, P. M., July 20, 1898, but it was not found until March, 1906. The meteor of July 20, 1898, seems to have traveled in a direction somewhat west of north, and its flight is said to have been accompanied by a heavy rumbling noise and a "trail of fire ten or twelve feet long." The meteorite was found about two miles north-northwest of Selma, Alabama, near the road to Summerfield, and it takes its name from the nearest town, as is the rule with meteoritic falls.

Selma weighs 306 pounds in its present condition, and it is probable that its original weight was about four pounds more, one or more small fragments having been lost from the mass. It is one of the ten largest aërolites ever found, and is the fourth largest aërolite that has fallen in the United States. The others having been broken up, this is probably the largest entire stone meteorite in the country at the present time. Its dimensions are: length, as it rests on its pedestal, 20½ inches; width, 20 inches; height, 14 inches.

In shape Selma is roughly polyhedral without pronounced orientation features, but it is probable that the upper side, as the specimen now lies, was the "brustseite" or front during the flight of the mass through the atmosphere. This side is bluntly pyramidal in shape. The original glassy crust of the meteorite has been mostly decomposed and washed away so that the characteristic thumb-marks, or piëzoglyphs, have been partly obscured. These peculiar markings may be seen on the front of the meteorite and in the illustration on page 37. The mass is deeply penetrated by cracks on both sides, and the position and character of the fissures indicate that they were caused by unequal heating during flight through the atmosphere, the tension produced not being enough, however, to cause complete fracture.

During the years while the meteorite lay buried in the ground alteration due to decomposition advanced considerably. A cut and polished fragment shows the unaltered stone to have a dark brownish-gray color and to be made up of spheroidal "chondrules" firmly imbedded in a matrix of similar matter. The largest chondrules observed are one eighth inch across, but these are extremely rare and most of the particles have less than half this diameter. Close examination with a strong magnifying glass enables one to see minute grains of iron scattered through the mass. The stony portion of this meteorite consists of olivine, enstatite and a monoclinic pyroxene, while the iron contains a little troilite. The specific gravity is 3.42.